3.00

SO-EJI-426

Atom, Man,
and the Universe

Atom, Man, and the Universe
The Long Chain of Complications

Hannes Alfvén
The Royal Institute of Technology, Stockholm

Translated by John M. Hoberman

W. H. Freeman and Company
San Francisco

This book was first published in Stockholm, in 1964, under the title *Atomen människan universum* by Bokförlaget Aldus/Bonniers.

Copyright © 1969 by W. H. Freeman and Company.

No part of this book may be reproduced by any mechanical, photographic, or electronic process, or in the form of a phonographic recording, nor may it be stored in a retrieval system, transmitted, or otherwise copied for public or private use without written permission from the publisher.

Printed in the United States of America.

Library of Congress Catalog Card Number: 69-15872

*Would you that spangle of Existence spend
About* THE SECRET—*quick about it, Friend!
A Hair perhaps divides the False and True—
And upon what, prithee, may life depend?*

THE RUBAIYAT OF OMAR KHAYYAM

Contents

I

HOW NATURAL SCIENCE WORKS 7

II

THE LONG CHAIN OF COMPLICATIONS 13
The three fronts of natural science 13
Elementary particles 14
Protons and neutrons build atomic nuclei 21
Atomic nuclei and electrons build atoms 26
Atoms build molecules and crystals 34
Molecules build cells 40
Cells build plants and animals 45
The formation of societies 50

III

ATOMS AND MEN 55
The sense organs as physical instruments 55
The nervous system and the electric impulse technique 59
Vision and television 63
Mathematics and machines 65
The computer 68
A new link in the long chain of complications? 73

IV

THE COSMIC PERSPECTIVE 75
The creation 75
Galaxies and stars 81
Planets and satellites 83
The origin of the moon 86
Are we alone in the universe? 91

V

NATURAL SCIENCE AND SCIENCE 97
The philosophy of natural science 103

I
How Natural Science Works

Man has a deep-rooted impulse to relate himself to the world he inhabits. Not only does he ask how the world is constructed, how it was created, and what will become of it, but, more important, what is his own position in the world's Grand Design? Throughout human history, he has attempted to answer these questions, but since he was not present at the creation, cannot see into the future, and has not yet conquered outer space, his total knowledge, or assumed knowledge, of the world has been limited to speculation. This speculation is necessarily based on observation.

Man's first and most primitive assumptions about his world were based upon accidental observations on his immediate environment, which were more or less arbitrarily combined into an often elaborate fantasy. For example, the theory on the origin of the earth and of man that we are most familiar with is that in the book of Genesis. It was obviously the prevalent view at the time and place in which the book was written, and it was therefore incorporated into it. Because of the strong religious and cultural significance that the Bible still exerts, this assumption continues to be important, though our expanded knowledge has long since made it appear unreasonable. The systematic collection and treatment of observations, which we call natural science, has provided us with a basis for our speculations about the world that is entirely different from that of several thousand years ago. In this book we will discuss those results of natural science that are most important in this respect.

But, first of all, we shall examine the ways in which natural science works and the goals it sets up for itself.

The great industrial revolution that started in the Western world during the nineteenth century is in great part the result of natural science. The science of electricity has given us electric lights, electric motors, the telephone, radio, and television; chemistry has contributed a great variety of new materials; biology has created new medical techniques and improved grains. The fruits of science have also been used for inventing increasingly horrible means of destruction. All this has led many people to believe that the purpose of science is mainly technological advancement—that the task of science should be to produce better television sets, more durable nylon stockings, and more effective atomic bombs.

This is a misrepresentation. The goal of natural science is first and foremost to satisfy human curiosity by finding out how the world around us actually looks and by bringing order into our chaos of experiences and observations. It is true that what we learn about nature makes it possible for us to use it and master it, but this is not the primary objective of science. For example, the prerequisites for inventing a radio were a knowledge of the laws of electromagnetism, the discovery of radio waves, and an understanding of how the properties of electrons were made use of in a transistor. But Maxwell did not set forth the laws of electromagnetism with any thought of their practical application; and neither Hertz, who discovered radio waves, nor Thomson, who discovered the electron, dreamed of what their work would one day make possible. To take another example, much of the progress in nuclear physics was made during the years between the two world wars. Although it was this progress that created the scientific conditions necessary for the atomic bomb, none of those who carried out the work had any thought of such a result. It was not until they felt the impact of the Second World War that they reluctantly used their knowledge in the service of war technology.

The conversion of science to technology was the result of inventive activity and, to a larger degree, of its systematic derivative, which we call applied research. In spite of the admitted practical significance of applied research, we shall nevertheless devote what follows exclusively to the "impractical" side of science.

The great volume of scientific work has made radical specialization necessary. A chemist, an astronomer, or a botanist does not understand much about the others' fields. However, specialization is being counteracted to some degree by the increasing interest in the fruitful border areas between the different sciences that have been opened up in recent times. Thus astronomy and physics have been combined into astrophysics: ever since physicists discovered the relationships that exist between the spectrum generated by a light source and the properties of this light source, astronomers have been able to draw very important conclusions about the composition of the stars by analyzing the spectra of starlight. Similarly, a combination of physics and chemistry has yielded physical chemistry; and the application of chemistry to biological problems has initiated the extremely fruitful field of research that we call biochemistry.

But even within each scientific discipline, <u>specialization has resulted from the necessity of applying many different methods to a single subject. This specialization has produced three different types of scientists.</u>

The first group includes the sample collectors and the systematizers. They investigate flowers, birds, insects, or rocks, which they catalog; they analyze and synthesize known and unknown chemical compounds; they count stars and classify them; or they make precision measurements of spectral lines and calculate the energy states of atoms. It is this group that is responsible for the traditional image of scientist: incredibly diligent, precise, absorbed in his work and therefore quite introverted. This is the group that constructs the solid basis for all of science.

The second group might be characterized as the engineers of science. Their task is to invent and construct the increasingly complicated instruments that science requires. They are the men obsessed with records: for them scientific progress is measured by the maximum pressure of the highest temperature that can be achieved, by the resolution of the latest giant telescope, or by the particle energy attained in the most modern accelerator. It is they who widen the limits of science and make it possible to study ever more remote star systems, or particles with yet shorter lives. They send up the satellites and design the spaceships. Because their activity is extremely expensive, funds must be acquired; and if the work is to be useful to science, the projects selected must be truly important. Since publicity is essential to the acquisition of funds, this type of scientist has, of the three, become the most prominent in the public eye. Their rise to the fore is naturally aided by the fact that many people are much more easily impressed by the world's largest, most majestic, and most beautifully built telescope than by the characteristics of insignificant stars that the telescope is built to investigate.

The theoreticians form the third group. Their function is to treat the results obtained by the first two groups, expressing them in as clear and precise a form as possible: in other words, to construct a theory. To them the aim of science is to summarize as much experience as possible, to demonstrate that even the most widely disparate occurrences may be essentially similar and merely unlike aspects of one fundamental phenomenon. Although the names of the great theoreticians are well known, not everyone understands the way in which they work. Some of their work is related to artistic activity: for both the artist and the scientist sort the essential from the chaos of sensory impressions, and render this in as concentrated and elegant a form as possible. Just as the painter expresses his thoughts and experiences in colors, the sculptor in clay, and the musician in notes, so the practitioners of the art of science use formulas and laws that—like anything that offers a concentrate of the world we live in—exhibit a high degree of beauty. The highest praise a theoretician can receive, when he shows a new formula to a colleague, is the enthusiastic cry, "Very beautiful!" In reality the beauty of the formula differs no more from that of music than the beauty of music does from that of painting. It is true that the vision of science as art is a most exclusive experience, and is enjoyable only after many long years of study; but a correct interpretation of an atonal symphony or a cubist painting also requires a certain amount of preparation—the taste must first be cultivated and specialized in a given direction. The Greeks numbered astronomy among the fine arts, and its Muse was Urania. The other natural sciences were not included because they were not yet in existence at the time that Mnemosyne's nine famous daughters were born.

Although it is obviously impossible to describe an activity as variegated and richly faceted as science in one small paragraph, we can say that scientific work takes place in the following way: When investigation of an area of research begins —whether the field has been known for a long time or whether it is a new one initiated by a series of recent discoveries—attempts to anticipate which laws might apply within that area are quickly made. Hypotheses are propounded, and these are gradually formed into theories that are at least partly worked through. The theories are intended to summarize all the facts that have been found and even to predict the results of new investigations. If the predictions are subsequently verified a theory is "confirmed," but if it is not confirmed, it must be replaced by another theory. Not infrequently, two or more neighboring areas can be covered by a common theory, and it is consequently desirable to generalize theories so that they summarize all the results within as large an area as possible.

That <u>summary of experience constituting a theory must be formulated in a way that, although often abstract, is extremely concise. Science is therefore in need of a language that makes concentrated and logical formulation possible. Mathematics is such a language.</u> Mathematical formulas make it much easier to express a theory exactly, and with the help of mathematical methods one can analyze the content of a theory and predict its consequences. It has been suggested many times that man would not have been able to think logically if no language existed. Whether this is true or not, it is obvious that ordinary language does greatly facilitate organized thought. And the language of mathematics is an even greater aid to the formulation of scientific thought. The "mathematical apparatus," as the system of formulas and arithmetic laws that mathematicians have put at the disposal of natural science is called, is indispensable for unifying complicated arguments and deductions.

<u>Although all theories *can* be formulated in ordinary language, most of them would lack the sharpness and elegance that mathematics makes possible.</u> An entire book may be required to express in words that which is contained in half a line of formula. The translation of a mathematical formula into literary language is certainly more difficult than even the

translation of Chinese poetry, and the beauty of certain formulas is always lost.

One often hears the assertion that a theory is "mathematically proven." This expression is misleading. It is equivalent to maintaining that there is a mathematical proof that grass is green. A theory is a summary of observations, and its validity, or lack of it, can be ascertained only by comparing it and its implications with observations. Mathematics is invaluable for making it possible to survey all the implications of a theory with certainty and clarity, but the final "proofs" of a theory's accuracy can be provided only by observations.

As an example of how natural science works we shall discuss some of the developments that took place in the history of physics. Although many important natural laws had been discovered by the Mediterranean, Indian, and Chinese philosophers, Galileo's discovery of the laws governing the motion of a falling object is considered by many to be the birth of modern physics. Of perhaps greater significance than the formulation of these laws were the new principles of scientific thinking that were introduced in the process. The most important objective for Galileo was not to find out *why* a stone fell, but *how* it fell—what were the laws explaining its increase in velocity, and what described the relationship between the height from which an object fell and the duration of the fall? In other words, he realized that it was not essential to determine the "ultimate cause" of an event, and he confined himself to a study of the event itself. As a result of this differentiation, a division between metaphysics and physics was made and it has existed ever since. The function of physics, then, as well as that of the other natural sciences, is the description and coordination of occurrences rather than the "explanation" of them. Science tries to relate to one another as many widely different phenomena as possible, to demonstrate that all are actually unlike aspects of one and the same thing; but this is not exactly the same as "understanding" these phenomena.

Astronomy evolved in much the same way as Galileo's mechanics, and after the introduction of the Copernican system, Kepler was able to set forth his famous laws of planetary motion summarizing a very large number of observations of the planets' motions in the heavens. In his analyses, Kepler relied so heavily on Tycho Brahe's extraordinarily precise (for that time) measurements of the planets' positions, that one might say his laws are a synthesis of all the measurements made by Tycho Brahe in the course of many years of clear nights.

Astronomy and the science of falling bodies, which had previously been separate disciplines, were combined by Newton, who showed that Galileo's law of falling motion and the planetary laws could be seen as special cases of much more general laws that were applicable to the motions of all bodies: for a stone dropped from a tower, for a meteorite falling toward the earth, for the planets that moved in the heavens. Newton's great synthesis, usually called *classical mechanics* today, was extended and deepened during the eighteenth century, revealing for the first time a large area within which all phenomena could be calculated in detail according to a single basic law. By applying this single law (which in mathematical symbols fills half a line) we can determine the motions of the moon and the planets in the sky, the place at which a hurled projectile will land, the height and magnitude of the waves created by a steamship, the tone and the sonority of a flute, or the maximum cargo of an airplane. The significant test, then, of classical mechanics has been the revelation that all these apparently widely disparate phenomena are indeed merely unlike aspects of the same phenomenon.

The science of electricity (electrodynamics), too, began as

two separate sciences: electrostatics, which treated those phenomena that occurred when a bit of amber or a glass rod became electrically charged by rubbing it; and magnetostatics, the study of magnets and magnetic fields. After Galvani and Volta had shown that electricity could be produced with chemical elements, and Ørsted subsequently found that the electrical current produced in this way had magnetic effects, these fields were combined. The founder of electrodynamics was Maxwell, whose famous equations summarized completely the extensive field as it existed in his time. But there was more to come. When Maxwell examined the implications of his equations he found that, among other things, they predicted a wave motion of an electromagnetic nature. In attempting to verify Maxwell's predictions, Hertz did find such waves (which we now call radiowaves). After it was demonstrated that light, too, was an electromagnetic wave motion, optics (the science of light) became a branch of electrodynamics. From the study of optics we have learned that such differing effects as the refraction of light in a lens or its reflection by a mirror, and the functioning of an electric motor or a television set, are all explained by the one natural law that is formulated in Maxwell's equations.

Both classical mechanics and electrodynamics were largely closed chapters by the end of the nineteenth century. With the advent of the twentieth century, a chaotic period for physics began. New discoveries made it possible to study atomic structures, and it was soon apparent that events taking place in the microworld of the atom were not governed by the same laws that applied to the phenomena that had been studied up to that time. The motions of the electrons, which circled the minuscule but heavy nucleus, did not conform to the laws of classical mechanics. Quantum mechanics (or wave mechanics), which was developed during the 1920's, supplied the answers to our questions about electrons, and as a result we are now well informed about the structure of the atom outside the nucleus.

Quantum mechanics can be considered a generalization of classical mechanics, or, conversely, classical mechanics can be regarded as a special case of quantum mechanics. As soon as we begin to work with such extremely "small" phenomena as the structures of atoms, we must employ quantum mechanics; but for calculating the motions of larger bodies, quantum mechanics always gives the same results as classical mechanics.

This, then, is how the scientist works. He first looks for those laws which apply within a certain area, and when he has found them he tries to extend them to new areas. Occasionally the laws can be applied without being changed, as were Maxwell's electromagnetic laws to light phenomena. At other times, the laws must be made more general before two areas can be combined, as were those of the field of mechanics when it was combined with atomic physics. It can be said that the ultimate goal of natural science is to discover a single law or formula that will explain all experiences and all observations. We do not know how long we will have to work, before this goal is attained—certainly a very long time. But we have already gone a good part of the way: this is evident in certain quite large and important areas, such as electrical science, in which all known phenomena have already been summed up in a single law.

Let us assume that we did find the "ultimate law" of nature we seek, so that we could proudly assert, "In this way the world is constructed." Immediately a new question is raised: what lies behind this law, *why* is the world constructed in just that way? This "why" leads us beyond the limits of natural science, and into the area of metaphysics or religion. A physicist, as an expert, should answer with an *ignorabimus:* we do not know and can never know. Others would say that God established this law when he created the universe. A pantheist would perhaps say that this ultimate law *is* God. We shall not attempt to decide which answer is most satisfactory, since this problem is outside the scope of natural science. If its solution is essential to a total knowledge of our world—and some would say that it is most essential—we must answer that natural science cannot provide this total knowledge, but only the facts on which it can be built.

As mentioned before, <u>it will be a long time before we find such a basic law of nature.</u> The modern giant accelerators have enabled us to discover a number of new elementary particles, but the higher the particle energies attained, the more confusing the phenomena that are discovered. Nuclear physics, too, is to some degree an unfinished chapter: we know enough to make atomic bombs, among other things, but our knowledge of the forces that hold the nucleus together is still incomplete. But except for this subject of nuclear forces, physics is, in principle, a finished chapter. We certainly cannot explain all of its phenomena by a single natural law; but we have explained them by a few simple—although abstract—laws. This achievement is of extraordinary significance, since this part of physics includes all of "everyday" physics—actually, almost everything we can observe without the aid of sophisticated instruments. The greater part of astronomy, all of chemistry, and much of biology belong, in principle, to this area, and can, in principle, be explained by these known physical laws. In the following chapters we shall see what is contained in the reservation "in principle."

II

The Long Chain of Complications

THE THREE FRONTS OF NATURAL SCIENCE

If we are to survey anything as extensive and diverse as modern natural science, we must generalize; and if we generalize, we run the risk of saying only a fraction of the truth. With this in mind we may state that the modern study of natural science is concentrated on three major fronts. We can classify them into the examination of the very *large*, that of the very *small*, and that of the very *complicated*. These are the three main fronts on which man combats his ignorance.

The study of the very *large* is astronomy. With more and more highly complicated instruments astronomers observe ever more remote objects and attempt with increasingly refined theoretical methods to form an understanding of how the world we inhabit looks in macrocosm. The astronomer focuses not only on the great distances, millions and billions of light-years, but also on long periods of time. How has this universe developed during millions and billions of years? What is it that we are in the process of reaching? Can we predict what will happen at a time in the distant future? And, most important, what is our own status in the universe? We know that in astronomical terms we are very, very small; but do we have comrades in this great universe as small as we, but just as daring in their attempts to analyze the astronomical infinities? If not, we are unique.

The *very small* is the world of the atoms. We and everything around us consist of atoms, and it is of primary interest to us to understand these building blocks of which we are composed. But these are not as simple as was previously believed: for each mystery solved by atomic physics, new and even deeper ones appear.

Astronomy and atomic physics, then, are complicated sciences that investigate the fundamental laws that govern both the universe and ourselves. But there is no doubt that the area of the *very complicated* belongs to biology. It is true that we are constructed of atoms, but even if we completely understood the properties of these atomic components, we would understand very little about ourselves. Our relationship to the world of the atoms is what we shall call the long chain of complications.

ELEMENTARY PARTICLES

The word atom means indivisible. It was introduced by the Greek philosophers to designate the tiniest particles that matter was thought to consist of. The physicists and chemists of the nineteenth century adopted the term and used it to denote the smallest particles known to them. The name has endured although we have long been able to "split" atoms, so that the indivisible is no longer indivisible. Today it is thought that the tiniest particles composing the atoms are what we call *elementary particles*. Other elementary particles, which are not actual constituents of atoms, also exist. The most common way to produce these is with the use of the large cyclotrons, synchrotrons, or other accelerators that are constructed especially for studying these particles. They are also produced when cosmic rays pass through the atmosphere. When these elementary particles are produced they disintegrate within a few millionths of a second, or often within an even much shorter period of time. Upon disintegration, either they are transmuted into other elementary particles or they emit their energy in the form of radiation.

The study of elementary particles is concentrated upon the increasing number of short-lived elementary particles. Although this subject is of great significance, particularly because of its relevance to the most fundamental laws of physics, the research has, at present, little contact with the other branches of physics. For this reason, we will limit the following discussion of elementary particles to those that are permanent components of our common materials, and to some of the particles most closely related to them.

The first of these to be discovered, at the end of the nineteenth century, was the electron, which has become an extremely useful servant. In the tubes of a radio, a stream of electrons moves in a vacuum, and it is the control of this stream that strengthens the incoming radio signals and converts them into sound—or noise. In a television set, a beam of electrons serves as a pen that instantaneously and accurately copies onto the receiver's screen what the sender's camera sees. In both of these examples, the electrons move in a vacuum so that their motion may be as undisturbed as possible. Another useful property of electrons is their ability to illuminate a gas as they flow through it. Thus, by allowing electrons to pass through glass tubes containing gas at a suitable pressure, we utilize this phenomenon in the neon lights that illuminate the entertainment areas of every large city at night. Yet another effect of electrons is seen when lightning strikes, and an enormous number of electrons crash their way through the air to produce the roaring sound of thunder.

However, under the conditions prevalent on earth, there are relatively few electrons that have the freedom of motion exhibited in the preceding examples. Most of them are securely bound up in atoms. Because the atomic nucleus is positively charged, it attracts electrons, which are negatively charged, and it confines them to orbits that are relatively close to it. An atom normally consists of a nucleus and a number of electrons. If an electron escapes from an atom, it is as a rule immediately replaced by another electron, which the atomic nucleus draws to itself with strong attractive force from its immediate environment.

In the atoms of metals (and of other electrical conductors) all the electrons are not bound as tightly as are those in the atoms of other substances. One or two electrons from each atom have some freedom of motion. The forces emanating from the atom do not prevent them from moving quite freely within the metal, but the combined forces of all the atoms in the metal make it difficult for them to escape from the metal in all but exceptional cases. As a consequence of this, a stream of electrons can be sent through a metal wire. The electrons move quite unhindered throughout the wire but cannot leave it, and thus they stream through the wire in the same way that water moves through a pipe. It is this phenomenon that has produced one of the most common landmarks of the modern world: the long metal wires and the poles that support them; through these wires or cables, electrons transmit messages or power.

But then what does this remarkable electron look like? No one has seen it, and no one ever will; but nevertheless we know its properties so well that we can predict in detail how it is going to behave in different situations. We know its mass (its "weight") and its electrical charge. We know that most often it behaves as though it were a very tiny *particle*, but that at other times it has the qualities of a *wave*. The very abstract but also very exact theory of the electron, brought to completion several decades ago by the English physicist Dirac, enables us to determine in which circumstances the electron will most resemble a particle, and in which its wave nature will prevail. This dualism—particle and wave—makes a clear picture of the electron difficult; consequently a theory that takes both concepts into consideration and yet gives a complete description of the electron must be very abstract. But it is also unreasonable to restrict the depiction of a phenomenon as remarkable as the electron to such mundane images as peas and waves.

One of the implications of Dirac's theory of the electron was that there must be an elementary particle that had the same properties as the electron except that its electrical charge was positive instead of negative. Indeed, such a counterpart of the electron has been found. It has been named the *positron*. It is a constituent of cosmic rays and it is generated by the decay of certain radioactive substances. Under earthly conditions, the positron has a short life. As soon as it comes into the neighborhood of an electron—and this happens in all substances—the electron and the positron "annihilate" each other. The positron's positive electric charge neutralizes the electron's negative charge. Since, according to the theory of relativity, mass is a form of energy, and since energy is indestructible, the energy represented by the combined masses of the electron and the positron has to be somehow preserved. This is done by a photon (a quantum of light), or usually two photons, which are emitted when the annihilating collision takes place; their energy is equal to the combined energy of the electron and the positron.

We know also that the reverse of this process takes place. A photon can, under certain circumstances—for example, when it passes near an atomic nucleus—produce "out of nothing" an electron and a positron. To do this, it must have at least as much energy as that corresponding to the combined masses of the electron and positron.

Elementary particles are not, then, eternal or permanent. Both electrons and positrons can be born and can die. But the energy and the resulting electrical charges are preserved.

With the exception of the electrons, the elementary particle that we have been aware of for a longer period of time than any other is not the rare positron, but the *proton,* the nucleus of the hydrogen atom. It is positively charged, as is the positron, but its mass is nearly two thousand times greater than that of the positron or the electron. Like these particles, the proton sometimes exhibits a wave motion, but only under very special circumstances. That its wave nature is less conspicuous is actually a direct consequence of its having a much larger mass. The wave nature, which is characteristic of all matter, does not become important to us until we begin to work with extremely light particles, such as electrons.

The proton is very common. A hydrogen atom consists of a proton as "nucleus," orbited by an electron. The proton is also a constituent of all other atomic nuclei.

Theoretical physicists had anticipated that the proton, like the electron, had a counterpart; the discovery of the *negative proton* or *antiproton,* having the same properties as the proton but a negative charge, fulfilled this expectation. If an antiproton collides with a proton, they are both "annihilated," as an electron and a positron annihilate each other.

Another elementary particle, the *neutron,* has almost the same mass as the proton, but it is electrically neutral (without electric charge). Its discovery in the 1930's—at about the same time as the positron—has been enormously important to nuclear physics. The neutron is a constituent of all atomic nuclei (except, of course, the ordinary hydrogen nucleus, which is simply a free proton), and when an atomic nucleus is broken, one or more neutrons are emitted. The explosion of an atomic bomb is effected by neutrons that are set free from uranium or plutonium nuclei.

Since protons and neutrons together make up atomic nuclei, they are both referred to as nucleons. After a certain period of time, a free neutron changes into a proton and an electron. Conversely, a proton can be changed into a neutron plus a positron.

We know of another particle called the *antineutron* which, like the neutron, is electrically neutral. It has many of the same properties as the neutron, but one of its major differences is that it disintegrates into an antiproton and an electron. If a neutron and an antineutron collide, they annihilate each other.

The *photon,* or light quantum, is an extremely interesting elementary particle. When we light a lamp to read a book, the lamp produces a very large number of photons, which rush toward the book and all parts of the room at the speed of light. Some of them are destroyed as soon as they strike a wall; others rebound several times against the walls or other objects, but, within less than a millionth of a second after they have been produced, all but a few that have escaped through a window and continued on out into space are destroyed. The energy required to produce the photons is supplied by the electrons that are sent through the lamp when the switch is turned on, and the photons leave this energy behind them when they are destroyed in a book, or other object, as they warm it, or in the eye, where they produce a stimulation of the ocular nerves.

A photon's energy—and consequently its mass—can vary a great deal. Thus there are very light photons as well as very heavy ones. The photons that constitute ordinary light are very light, having a mass only some millionths that of the electron.

Other photons have a mass about the same as that of an electron, and even much greater. X-rays and gamma rays are examples of heavy photons.

A general rule is that the lighter an elementary particle, the more pronounced its wave nature. The heaviest elementary particles, the protons, show comparatively slight wave characteristics; those of the electrons are somewhat more prominent; and the wave characteristics of photons are the most dominant of all. In fact, the wave nature of light was discovered long before its particle-like characteristics. We have known that light was an electromagnetic wave motion since Maxwell demonstrated this during the latter half of the nineteenth century, but it was Planck and Einstein who discovered, early in the twentieth century, that light did have particle characteristics, that it was sometimes emitted in distinct "quanta" or, in other words, as a stream of photons. It will not be denied that it is mentally difficult to fuse these two apparently unlike concepts of the nature of light; but we can say that, like the "double nature" of the electron, our concept of a phenomenon as subtle as light must be very abstract. It is only when we wish to express our conception in rough images that we must sometimes liken it to a stream of particles, photons, or to a wave motion of electromagnetic nature.

There is a relationship between a phenomenon's particle nature and wave properties. The heavier a particle, the shorter its corresponding wavelength; the longer the wavelength, the lighter the corresponding particle. The X-rays, which consist of very heavy photons, therefore have a very short wavelength. Red light, which has a longer wavelength than that of blue light, consists of photons that are lighter than those of the blue light. The longest electromagnetic waves of all, radio waves, consist of extremely tiny photons. Radio waves show no trace of particle properties whatsoever; their wave nature is the completely dominating characteristic.

The Long Chain of Complications

Finally, the tiniest of all the small elementary particles is the *neutrino*. It has no electrical charge, and if it has any mass at all it is somewhere around the vanishing point. With some exaggeration, we can say that it simply lacks properties.

Our knowledge of the elementary particles is the contemporary frontier of physics. The atom was discovered in the nineteenth century, and the scientists of that time found an increasing number of different kinds of atoms; similarly, today we are finding more and more elementary particles. Although the atoms were proved to consist of elementary particles, we cannot expect that, by analogy, the elementary particles will be found to consist of yet smaller particles. The problem we face today is different, and there is nothing to indicate that we will be able to split the elementary particles. We are more justified in expecting that they will all be shown to be manifestations of one even more fundamental phenomenon; and if this could be found, we would then be in a position to understand all the properties of elementary particles; we could calculate their masses and the ways in which they interact. Many approaches have already been made to this problem, which is one of the most important in physics.

PROTONS AND NEUTRONS BUILD ATOMIC NUCLEI

Three of the elementary particles—electrons, protons, and neutrons—make up all material. Since protons and neutrons change into each other easily and are both called nucleons, we might just as well say that matter consists of two building blocks: electrons and nucleons. Matter is built from these particles in two stages: in the first the nucleons combine into atomic *nuclei,* and in the second these atomic nuclei combine with electrons to form *atoms.*

An atomic nucleus consists of a number of nucleons that

have combined. This number varies from one to more than two hundred. The simplest atomic nucleus is the hydrogen nucleus, which consists of one free proton, and the most complicated of the normal atomic nuclei is the uranium nucleus, which contains 238 nucleons. All numbers between these two are also represented by the different atomic nuclei.

In attempting to explain how a number of nucleons can be held together as an atomic nucleus, we must assume that when two nucleons come very close to each other a particularly strong attraction between them takes place. The nature of this attraction is different from electrical attraction, which, for example, takes place between a positively charged proton and a negatively charged electron. The force of attraction between nucleons we call nuclear force, and we acknowledge that closer research into its properties is perhaps the single most important task facing nuclear physics.

In order to visualize the construction of an atomic nucleus it is helpful to think of the nucleons as small balls, which adhere to each other when they are very close together; that is to say, the nuclear forces hold them together in the form of a little, almost round, clump—an atomic nucleus.

The mass of an atomic nucleus is approximately the same as the combined mass of the nucleons of which it consists. For example, the nucleus of the iron atom, which contains 56 nucleons, is said to have an "atomic weight" of 56, and its mass is about 56 times the mass of a single nucleon. Actually, its total mass is somewhat less than 56 nucleon masses, because when these particles unite into a nucleus, a certain amount of energy, the so-called binding energy, is liberated and escapes; since all energy has mass, then, some mass has been lost as a consequence of the union of the nucleons. In all nuclei, however, the amount of mass lost is less than one percent of the total.

With the exception of the atomic weight, the most important characteristic of an atomic nucleus is its electrical charge. This determines the atom's chemical and most of its physical properties. The charge of atomic nuclei varies from 1 to about 100. The uranium nucleus, which has the greatest charge of all the naturally occurring substances, has a charge number ("atomic number") of 92, and other nuclei with yet higher atomic numbers, such as plutonium, have been artificially produced. The most frequently occurring uranium nucleus has an atomic weight of 238, thus consisting of 238 nucleons. Since protons have an electrical charge, but neutrons do not, we can say that, of the nucleons making up the uranium nucleus, 92 are protons and the rest are neutrons $(238 - 92 = 146)$.

If the atomic nuclei of two atoms have the same charge but different masses, they are *isotopes*. One nucleus, for example, has a charge of 92 and a mass of 235; this atom is thus an isotope of uranium 238. Since it is the charge that determines the chemical properties of the atom of which the nucleus is a constituent, both of the atoms having these isotopic nuclei have essentially the same chemical properties, and both are uranium. (The uranium isotope 235 is used for making atomic bombs.)

In order to explain in greater detail how nuclei are built from nucleons, we shall discuss some of the simplest nuclei. The simplest of all is the ordinary hydrogen nucleus, which is a free proton. If a neutron comes close enough to the proton, both particles can combine into an atomic nucleus consisting of a proton and a neutron. This nucleus then has a charge of 1 and a mass of 2. Since it has the same charge as an ordinary hydrogen nucleus, the atom is an isotope of hydrogen, called "heavy hydrogen," or deuterium. If another neutron is added, a third hydrogen isotope is obtained, "extra heavy hydrogen," or tritium, with an atomic nucleus having a charge of 1 and a mass of 3. If a proton instead of a neutron is added to the deuterium nucleus, the result is a nucleus with a mass of 3 but a charge of 2. Since a nucleus with a charge of 2 is a helium nucleus, we can change deuterium to helium by causing a deuterium nucleus to accept a proton. This atom is not, however, the most common helium isotope, which has a mass of 4, but a lighter isotope, called helium 3.

If we compare the tritium and helium 3 nuclei, each of which consists of three nucleons, we find that helium 3 has less energy than tritium, so that a conversion from helium 3 to tritium requires a supply of energy from the outside. If, on the contrary, tritium is converted to helium 3, energy is given off. This kind of conversion also occurs spontaneously: a quantity of tritium changes by itself (in the course of several decades) to helium 3. We call this *radioactivity*. The nucleus, which originally consisted of a proton and two neutrons, contains after the conversion two protons and a neutron; what has happened, then, is that a neutron has been changed into a proton. During this conversion, an electron is produced in the atomic nucleus in order to preserve the total electrical charge, and this electron is expelled from the nucleus at great velocity when the radioactive conversion is completed.

If an atomic nucleus consists of, say, 16 nucleons, about half of these must be protons if the nucleus is to be stable, so that it will not fall apart. The ordinary oxygen nucleus consists of 8 protons and 8 neutrons, and is stable. If the 16 nucleons consist, instead, of 7 protons and 9 neutrons, they form a nitrogen nucleus, which is unstable and changes radioactively into oxygen by changing a neutron into a proton and sending out an electron. A combination of 9 protons and 7 neutrons yields a fluorine nucleus, which also disintegrates into oxygen. But since this change results from the conversion of a proton to a neutron, a positron is emitted, taking with it the surplus positive charge.

As this example shows, only certain combinations of protons and neutrons yield stable atomic nuclei. All stable atomic nuclei, as well as some that are radioactive, are found in nature. A radioactive substance is unstable, so that sooner or later it disintegrates; but certain nuclei, such as that of normal uranium, decompose so slowly that only a fraction of such a nucleus has been lost since its formation early in the development of the universe. Radioactive substances that are more short-lived are also found in nature. Today we can artificially produce many hundreds of different radioactive substances.

Enormous quantities of energy are liberated during many nuclear reactions. When a substance disintegrates radioactively, a large amount of energy is produced as a result, but since all the radioactive substances that are available to us in large quantities disintegrate slowly, the release of the energy occurs in the course of such a long period of time that it is not very dramatic. It was not until we succeeded in splitting the uranium and the plutonium nuclei that we were able to attain that swift and intensive release of energy that constitutes the explosion of an atomic bomb. Another and incomparably more important atomic reaction takes place in the interior regions of the sun and the other stars, supplying them with the energy they then emit into space. This reaction is rather complicated, but the result is that four protons combine into a helium nucleus and emit two positrons. The sun's hydrogen is thus gradually "burned" into helium. Without the heat from this fire the earth's temperature would soon sink toward "absolute zero," or 273 degrees centigrade below zero. Man has not yet been able to produce in large scale this atomic reaction, which is much more effective than the fission of uranium for the release of energy, but it is not improbable that we shall soon succeed in utilizing it or a similar process (thermonuclear power).

The atomic nuclei, which together with electrons make up the world we inhabit, were probably formed several billion years ago by the combination of free protons and neutrons.

It is possible that this process took place in the interior regions of the stars.

At present large-scale nuclear reactions occur within the sun and the stars. The temperature at the sun's center is around 20 million degrees, and this is just sufficient for "igniting" hydrogen so that it will burn to helium. These reactions produce a large number of neutrons, and if several of these attach themselves to protons, heavier elements are thereby produced. In certain stars, whose interiors are very hot, nuclear processes are very effective; in exploding stars ("novae" or "supernovae") in particular, a very substantial production of heavier elements can be expected. There is thus a possibility that elements are built up within the stars and afterwards expelled into space.

These, then, are some of the more sensational aspects of nuclear physics, but there remain problems that, if more commonplace, are no less interesting or important.

ATOMIC NUCLEI AND ELECTRONS BUILD ATOMS

Since positive and negative electrical charges attract each other, an atomic nucleus, which is always positively charged, will attract the negatively charged electrons. An electron that is attracted to a nucleus does not become part of the nucleus but orbits around it at a certain distance. The reason for this is that the electron behaves as a wave rather than a particle. When a hydrogen nucleus having only a single unit of positive charge has captured an electron in this way, a hydrogen atom has been formed. Outwardly, this atom is electrically neutral, since the electron's negative charge neutralizes the positive charge of the nucleus. A helium nucleus, which has a double charge, must attract two electrons to form a neutral atom, the nucleus of iron (having an atomic charge of 26) requires 26 electrons, and finally the uranium nucleus (atomic charge, 92) normally surrounds itself with 92 electrons.*

The atomic theory put forth by Niels Bohr in 1913 included the first conception of how the atom is constructed. According to this theory, electrons move in circles or ellipses around the nucleus, in somewhat the way that planets move around the sun. There is, however, a large difference between the motion of the planets and that of the electrons. A planet can move at any distance from the sun. Thus if the speed of the earth's revolution around the sun were diminished, the earth would fall a certain distance toward the sun to move in a new orbit whose distance from the sun would be completely determined by the degree of deceleration. By either decelerating or accelerating the earth's progress, one could cause it to orbit in any ellipse (the only restriction being that a focus of the elliptic orbit lie in the sun). The electrons in an atom do not have such great freedom to choose an orbital path. The reason is that the electrons behave as waves which must form certain wave patterns.

As we mentioned earlier, the lights of a neon tube are caused by passing an electric stream through the inert gas, neon, which is contained in the tube. The neon nucleus has an atomic charge of 10 and is therefore normally surrounded by 10 electrons. If an electrical current is passed through the gas, this means that electrons are forced to move through it, and these collide with the atoms of the gas. If such a collision is powerful enough an electron can be broken loose from the

---

* In principle, an atomic nucleus can be built also from antiprotons and antineutrons. Such a negatively charged nucleus can be surrounded by positrons, and the resultant "antiatom" will have precisely the same properties as an ordinary atom. Such antiatoms can form "antimatter" having the same properties as ordinary matter. If matter and antimatter come in contact with each other, they annihilate one another, releasing an enormous amount of energy. According to certain theories a part of the universe consists of antimatter.

neon atom, and the remaining body is called a positive *ion*. A neon ion, then, would consist of a neon nucleus and nine electrons. The detached electron helps to carry the electrical current through the gas, but it can eventually be captured by another neon ion. Of course, the reason the ions are positively charged is that their nine negative electrons do not completely neutralize the ten positive units of charge within the nucleus, and the electrical field from an ion attracts neighboring electrons.

Only the collisions that are unusually powerful lead to the ionization of an atom. But frequently, as a result of a collision, one of the neon atom's electrons is hurled from its normal orbit to a new orbit at a greater distance from the nucleus. This is called *excitation* of the atom. The atom is still electrically neutral and therefore has, if it is neon, 10 electrons, although one of these has acquired an unusually high amount of energy because its distance from the nucleus has increased. Within a very short time it falls back to its normal orbit and its surplus energy is emitted in the form of a photon, which is produced when the electron jumps from one orbit to another. The light sent out by the gas when the electrical current passes through it consists of the photons emitted by the excited atoms when they return to their normal state. The photon's energy, and consequently that of the emitted light, is determined by the energy difference between the two orbits. If this is very small, the light is red and is emitted in long waves, a somewhat larger energy difference produces yellow and green light in shorter wavelengths, and yet a greater difference yields blue or violet light having an even shorter wavelength. By measuring the wavelength of the emitted light with precision, we can acquire worthwhile knowledge about the different energy states the atom can have. The clear red light sent out by a neon lamp is intended to tell the tooth-

The Long Chain of Complications 29

paste-buying public which brand to choose; but to the physicist it talks, instead, of the atomic construction of neon.

The different orbits that an electron can pass to during excitation of the atom, then, are of many different energy states. The light that is emitted by each passage between two orbits is called a *spectral line,* each of which has its particular color. Atoms of complex composition can emit thousands, or tens of thousands, of different spectral lines, and naturally it is difficult to analyze anything that complicated. In order to understand how an atom is constructed, it is therefore necessary to begin with an analysis of the simplest atom, the hydrogen atom, in which the nucleus is circled by only one electron. Normally this electron moves in an orbit very close to the nucleus (the orbital diameter is about one ten-millionth of a millimeter). By analyzing the spectrum emitted by hydrogen gas when an electrical current is passed through it, we can calculate the orbits the electron moves in when the atom is excited. As a result we have learned that the distance of the orbit from the nucleus can be 4, 9, 16, or 25 times the normal distance between the nucleus and the orbit, but it cannot be, say, 5, 8, or 13 times that distance.

It was at first difficult to understand why the electron could move only in these precise orbits and not, like a planet revolving around the sun, at any distance at all from the nucleus. The solution to this mystery was perceived when the wave nature of the electron was discovered. As we observed in the section on elementary particles, an electron sometimes resembles a particle but it has other properties that can cause it to behave like a wave. When we are working with extremely small dimensions, like those of the atoms, the wave properties are as prominent as the particle properties, if not more so, and it is not entirely accurate to say that the electron moves in a definite orbit. Instead, an electron moving around an atomic nucleus might be conceived of as an electrical charge that pulses, or vibrates, as it circles the nucleus. The problem of calculating how an electron can move in an atom is therefore not unlike the acoustical questions, how can a string vibrate, or which tones will a whistle produce.

A string on a violin or a piano can vibrate not only to its fundamental tone but also to a series of overtones. If one presses loosely on the string's midpoint the fundamental tone is suppressed and the string can be made to give only the first overtone, which is an octave higher than the fundamental tone. We can also produce the second overtone, which is a fifth higher than the first overtone, the third, which would be two full octaves above the fundamental tone, and so on. A flute player changes from one octave to the next higher one simply by changing the applied airstream, thereby causing the flute to give the first overtone. Analogously, the charge of the electron in a hydrogen atom vibrates to the fundamental tone when the atom is in its normal state; but when the atom is excited, the electron has begun vibrating to one of the many possible overtones. At this vibration state the greatest part of the electrical charge is situated at a greater distance from the nucleus than it is at the basic state. The charge is concentrated mainly at that orbit in which, according to Bohr's theory, the electron should move during that particular degree of excitation.

The discovery of the electron's wave nature gave rise to a refinement of Newtonian classical mechanics, which we call *quantum mechanics* or *wave mechanics*. This subject centers on the electron's wave nature, whereas classical mechanics is concerned with only its particle nature. For large bodies, such as planets, or even meteors and hurled projectiles, a knowledge of the particle nature is sufficient, because if the bodies are large enough their wave nature is not at all noticeable.

The Long Chain of Complications

However, if we are studying particles as small as electrons, and dimensions as small as the atom's, the wave nature manifests itself so strongly that we must employ quantum mechanics.

With the development of this new field a complete description of all the properties of the hydrogen atom unfolded. Unfortunately, the model of the atom provided by quantum mechanics is extremely abstract, and it is impossible to obtain a graphic picture of it because the subtle and unusual phenomena that take place within the atom simply cannot be described by analogies from everyday life. If we are going to attempt a pictorial description of the atom, we should be prepared to face the fact that at least in certain regards it will be very misleading. If we emphasize the particle nature of the electron, we arrive at Bohr's model of the atom, in which the electrons are regarded as particles that move in certain definite orbital paths and that are capable of jumping from one orbit to another, emitting a photon in the process. Alternately, we can put the emphasis on the electron's wave nature and consider the atom's different states as the fundamental tone and the overtones of the electron's vibrations.

We require of an atomic theory that it explain all we know about the atom. This is a challenging demand, because not only are we familiar with the atom's many properties but, because of the extraordinary precision made possible by the spectroscope, we can determine an atom's many different energy states with amazing exactitude (often to less than a millionth of a part).

To begin with, let us consider the hydrogen atom. If we make a theoretical calculation of its properties, using quantum mechanics as a point of departure, and compare the results with what is measured experimentally, we find that theory and experiment agree so closely that there is no noticeable discrepancy. The present theory, then, gives an excellent summary of everything we know about the hydrogen atom, and, as we have seen, this is an extensive and particularly precise body of knowledge.

But if we take the next simplest atom, helium, the problem becomes more complex. The helium nucleus is circled by two electrons, and each electron is affected not only by the electrical attraction of the nucleus but also by the repulsion from the other electron. The fact that the electrons disturb each other's motions makes the situation very complicated. The application of quantum mechanics to the helium atom is therefore much more difficult than its application to hydrogen, and only after long and tedious calculations have we been able to find the theoretical energy values and thereby to calculate the spectrum of helium. By comparing the theory's results and the experimentally measured spectrum we find for the helium atom, too, the same convincing agreement that we found for the hydrogen atom.

If the theory holds for atoms having one and two electrons, it might be reasonably expected that it would also hold for atoms having many electrons. However, the more electrons there are, the more difficult is the problem of calculating the energy states, until the task of calculation becomes so enormous that no one can do it. Only for some of the simplest atoms have the calculations been completed with a high degree of precision. Also, calculations for certain energy states of the heavier atoms can be carried out with a reasonable expenditure of effort. In all these cases the theory agrees with the observations as well as one could reasonably wish.

We might say of the quantum-mechanical atomic theory, then, that *in principle* it summarizes all we know about the atom (outside the nucleus). Should a doubter announce himself and declare that he does not believe that the theory applies to, say, the iron atom, we could not present binding proof that it does. The iron atom has 26 electrons and each of these is affected not only by the nucleus but also by its 25 comrades. Many thousands of spectral lines have been measured in the spectrum of iron. The procedure for carrying out the calculations for a complicated atom is not difficult to describe in principle, but the life of a mathematician is much too short and the capacity of the computer too small for the calculations to be made precisely. Therefore the only way we can answer the doubter is to say that the theory applies to all those atoms for which we have been able to make exact comparisons, and that we believe that it is applicable also to the iron atom because we can see no reason why it should not be. As long as the doubter cannot give any weighty reason for the contrary hypothesis, it is incumbent upon him to carry out the calculations and show that they do not agree with the measured results before he can be taken seriously. But then the life of the doubter is not that long either. And the iron atom, with its 26 electrons, is not an unusually complicated atom. The uranium atom has 92!

If we heat a piece of iron, it melts at about 1500 degrees and boils into a gaseous form at about 2500 degrees. Heating causes the motions of the atoms to move faster; for the melting, as well as the boiling, to take place, the energy of motion must be great enough to enable the atoms to tear themselves loose from the forces that bind them together to form a solid body or a liquid. Iron gas consists of free iron atoms that move in straight lines and collide with each other. They can be likened to elastic rubber balls that rebound against each other upon colliding but thereupon move in a rectilinear fashion. If the velocity of the atoms diminishes through cooling, the substance changes to liquid or solid form. If, on the other hand, the temperature is increased, the collisions become so violent that the atoms break each other apart. At a temperature of 5,000 to 10,000 degrees the collisions between the atoms of iron gas tear single electrons from individual atoms, which thereby become ionized: the atom has released an electron, and what remains (a nucleus and 25 electrons) is a positive ion. At a yet higher temperature the atom loses another electron and thereby becomes doubly ionized. However, not all atoms are minus a like number of electrons at a given temperature; as some are losing electrons upon collision, others are capturing them, and as a result a fraction of the atoms are neutral, a fraction are singly ionized, another fraction are doubly ionized, and so on. At those low temperatures that we call "normal" the probability of an ionization occurring is small, but if it does accidentally occur, the atomic nucleus soon captures another electron. Therefore we regard it as "normal" that an atomic nucleus retains a certain number of electrons. On the other hand, in the interior of the sun where the temperature is more than 10,000,000 degrees the nuclei and electrons are "normally" free.

ATOMS BUILD MOLECULES AND CRYSTALS

Quantum-mechanical atomic theory is correctly regarded as one of the greatest triumphs of physics. The fact that it gives such an exact description of the atom's properties has given a firm basis, if not to all of natural science, to a very large part of it; for matter is composed of atoms. Once we know the properties of atoms, we are in a position to calculate how they are joined together and by which laws the construction occurs. With quantum mechanics as an aid, then, we can determine not only the properties of the atoms but also the properties of all the substances they make up: we can understand why gold is yellow and steel is hard, why hydrogen and oxygen combine to form water, and what happens when water freezes into ice or boils to steam. The problem becomes one of calculating in detail how the forces from many atoms work together in a way that is often complicated beyond comprehension; but if we know the basic laws of their composition we have mastered the rules of the game.

The fact that some of the problems of nuclear physics and particle physics are not yet solved is not really significant for natural science in general. Nuclear physics is rather sharply demarcated from the other branches of physics. It is almost exclusively the nuclear charge—and to a certain degree the mass—that is of importance to the structure of atoms. The inner structure of the nucleus, which perhaps is not yet entirely clear, becomes relevant only to certain extraordinary phenomena, such as the disintegration of radioactive elements and the explosion of atomic bombs. But all of the more "normal" occurrences can be traced back to the properties of the atomic shell; and these we have derived from quantum mechanics.

We have seen that an atomic nucleus is normally surrounded by a number of electrons equal to its units of charge. This number of electrons is required to neutralize the nuclear charge. If, in one way or another, the atom loses an electron, the remaining electrons are not enough to neutralize entirely the positive charge of the nucleus; in fact the effect of this positive charge extends beyond the atom itself, and as soon as an electron comes into the vicinity of the atom it is attracted to it as a possible substitute for the lost electron.

But even if the nucleus has surrounded itself with the right number of electrons, certain forces from the atom affect its surroundings. The strength of these forces is determined by the effectiveness with which the electrons screen the nuclear field and the symmetry of their distribution around the nucleus. In an atom the electrons have a tendency to form "shells." Each shell contains a certain number of electrons, and if it is fully occupied the effects reaching beyond the electrons in the shell are very small. The innermost shell, the "K shell," contains 2 electrons, which orbit very close to the nucleus. If an atom contains more than 2 electrons the additional electrons must circle at a greater distance from the nucleus. There is room for as many as 8 electrons in the "L shell," but if the number of electrons in an atom is greater than 10, the others must move in orbits that are even more distant from the nucleus and that form the "M shell." If the number of electrons contained in an atom of a certain substance is not enough to fill a shell entirely, the electron distribution becomes asymmetrical, causing very strong forces to penetrate into the atom's environs. Forces of this kind are what hold two or more atoms together to form a molecule. An atom's chemical properties, that is, its ability to combine with other atoms to form a fairly complicated molecule, depend therefore on the structure of the electron shell. The forces that hold atoms together to form a solid are also of this kind.

If the outermost shell of a particular atom is complete, or "closed," the force field outside the atom, then, is very slight. Thus those atoms having a nuclear charge—and therefore an equivalent number of electrons—of 2, 10, and 18, for example, emanate weak force fields. All these substances are inert gases: helium with 2 electrons in the K shell; neon with 2 electrons in the K shell and 8 electrons in the L shell; and argon with 2 in the K shell, 8 in the L shell, and 8 in the M shell. The forces with which such an atom can affect other atoms are so weak that a chemical union is impossible. Thus the inert gases derive their name from the fact that they are indeed "inert," that is, unable to combine. Their atoms "prefer" to be free and independent, or in other words they constitute a gas. Only at very low temperatures can the inert gases be condensed into liquids or solids; but the forces that hold the atoms together are so weak that after only a slight warming, the resultant motion is sufficient to break the atoms loose from each other so that the substance again takes the form of a gas.

The force fields around the atoms of all substances other than the inert gases are so strong that the atoms can combine with other atoms to form molecules. The simplest example of molecule formation is the combination of two hydrogen atoms to form a hydrogen molecule. Such a molecule consists of two atomic nuclei (protons) and two electrons which, so to speak, hold them together. The two electrons form a K shell, causing the forces that emanate from a hydrogen molecule to be very weak. Thus, at normal temperature hydrogen is a gas, since the forces between the molecules are not strong enough to hold them together as a liquid or solid.

If, however, two carbon atoms are joined, the forces emanating from them are still so strong that several carbon atoms are attracted. The result is that a very large number of carbon atoms can join together in clumps. This cohesion can occur in an irregular fashion; or it can form a highly symmetrical structure, a crystal. Carbon in the form of a crystal is called a diamond. The atoms in a crystal are arranged in straight rows, and this perfect alignment can occasionally be preserved even in crystals as large as the Koh-i-noor diamond. The forces acting between the neighboring atoms determine the crystal's hardness and internal cohesive force. Metals also form crystals, but most of these are microscopically small, and a piece of metal consists of a very large number of small crystals that are baked together.

Different kinds of atoms often combine into molecules. Both sodium and chlorine are difficult substances to produce in a purified form, but their chemical combination, sodium chloride (or regular salt), is a very common substance. The sodium atom has 11 electrons, 2 in the K shell, 8 in the L shell, and the eleventh in the M shell. Chlorine has 17 electrons, 7 of the M shell's places being occupied. The sodium atom releases its most remote electron so that those remaining are in completely filled shells, and the chlorine atom readily accepts an electron in order to complete its M shell. If, therefore, sodium and chlorine are brought together, each chlorine atom draws to itself an electron from a sodium atom, causing the chlorine atoms and the sodium atoms to combine chemically with each other to form sodium chloride.

In a similar way an oxygen atom and two hydrogen atoms can combine to form water. The oxygen atom needs two electrons to form a completed shell and takes these electrons from two hydrogen atoms. Ammonia (NH_3) is an example of a molecule consisting of four atoms: one nitrogen atom bound to three hydrogen atoms. The sulfuric acid molecule (H_2SO_4) consists of seven atoms. Among the organic substances are examples of molecules containing thousands, and even millions, of atoms.

Of all the chemical combinations that atoms can form, the most interesting are without a doubt the organic substances. Their most important element is carbon, which usually exists in combination with oxygen and hydrogen. Many also contain nitrogen, as well as many other elements. It is not the size of the molecules that makes the organic substances remarkable. A crystal, for example, can be looked upon as merely a giant molecule, and a diamond contains an incomparably greater number of carbon atoms than any organic molecule. But in a diamond the atoms are arranged into a monotonously regular pattern—their alignment is straighter than the rows of a company of soldiers on parade. But the organic substances have a varied and changing structure. Carbon atoms can join together to form rings as well as long chains, which are sometimes branched, and to this carbon framework oxygen, hydrogen, and many other kinds of atoms are bound. It is the perfect symmetry that gives the diamond its hardness and its brilliance; but it is the irregularity, and the variety of combinations, that make the organic substances much more valuable than diamonds, for they are the bearers of life. Whereas a diamond has a stable, permanently given structure, the organic substances are transitory. A long carbon chain can be easily broken apart, but it can also be easily lengthened. Atoms can change their location within the molecule, and a particular organic combination changes readily into another one. This flexibility and variety of changing combinations is what makes it possible for the organic molecules to build up the unbelievably complicated structures of living beings.

All substances—be they air, water, earth, steel, glass, wood, or proteins—are built of atoms. Because our knowledge of atomic structure is detailed, we can calculate the forces with which the atoms affect each other. The chemical forces can be calculated in accordance with quantum-mechanical atomic theory, and from this we can also deduce the properties of all substances—at least in principle. As we have already mentioned, this reservation "in principle" is very important. We have seen how one can theoretically calculate the spectra of the simplest atoms and other properties with the same precision with which they can be measured. If we proceed to more complicated atoms, there is no reason why we should not be able to evaluate them theoretically; but in practice the work required to carry out such a calculation is discouragingly large. This limitation applies to chemical forces. From our knowledge of the hydrogen atom's structure we have calculated in detail the circumstances in which two hydrogen atoms combine to form a hydrogen molecule, and we have also calculated this molecule's different properties. The result of such a calculation concurs with what has been observed. For more complicated combinations the calculations have been necessarily rough. No discrepancy has been found between theory and observation, and there is no reason to expect any. *In principle,* therefore, all of chemistry can be said to concur with atomic theory, and the properties of all substances can be theoretically deduced from the fundamental laws of quantum mechanics.

In practice, however, the calculations for most substances become too complicated. It is simpler to investigate a substance's properties with the use of chemistry's ordinary methods than it is to figure them out theoretically.

Although this is especially true of the most complicated organic substances, the use of quantum mechanics is nevertheless extremely important in organic chemistry. With the help of quantum mechanics scientists have obtained many results that have a significant and direct bearing on organic chemistry and biochemistry.

During its infancy, chemistry was divided into inorganic and organic chemistry. It was believed, then, that whereas the inorganic substances could be synthesized by ordinary chemical methods, the organic substances could be formed only under the influence of a special force, the so-called "life force." But it is now obvious that no such boundary exists, and an increasing number of organic substances have been synthesized. To be sure, there are many organic substances that we have not succeeded in synthesizing, and that are produced only by life processes; but all laboratory results indicate that this failure is due only to the fact that such substances are so complicated. Chemical reactions as complicated as those that take place inside a cell cannot yet be induced in controlled circumstances in the test tube. But this does not mean that we have any reason to suppose that these processes are of any character other than that of ordinary chemical reactions.

MOLECULES BUILD CELLS

Let us review what we have observed thus far in this chain of complications we are studying. We have seen how two building blocks as simple as atomic nuclei and electrons form atoms with very complicated properties. This complex result is both striking and self-evident: a number of simple building blocks are joined together to form not only their sum but also what is called their *combination*. It is not true that the whole is simply the sum of its parts, since the way in which the parts are arranged to form a "whole" is a decisive influence. An oxygen atom is not merely an atomic nucleus and 8 electrons, but an atomic nucleus surrounded by 2 electrons in an inner shell and 6 electrons in an outer shell; it is an organism that is able to absorb or emit certain distinct spectral lines, bind certain other atoms in different chemical combinations, and take part in a series of complicated reactions. But all of these new properties have appeared automatically. No new element has been introduced during the construction of the oxygen atom; rather, we may consider it a self-evident consequence of the constituents' properties. If they are combined, precisely this result must occur according to the laws of physics. But the result, the oxygen atom, with all its complicated properties, is still something unexpected, something that our deductions could not predict. A combination of simple component parts has produced a new part with innumerable new properties.

In the next stage the atoms are the building blocks, which make up molecules; and in this step, too, new properties appear. As we have seen, ordinary salt, sodium chloride, consists of sodium, an easily oxidized light metal, and chlorine, a rather heavy green poisonous gas; it consists of these substances and nothing else. But their combination has properties entirely different from those of the component elements.

Now we are ready for the third link in the chain, the construction of the cell. The building blocks are neither electrons nor atoms, but extremely complicated molecules, largely protein molecules, and the result of their combination is a living cell, whose complexity exceeds the atom's as much as that of the protein molecule exceeds the electron's.

It should be pointed out, however, that this analogy between atomic and cellular composition is somewhat misleading. It is true that we can understand in detail how the simplest atoms, at least, are built of elementary particles, and that by combining such particles we can form an atom, but when we turn to the cell, even the simplest is so complicated that we have no truly detailed understanding of its construction. Consequently, there is little hope that we will be able to synthesize a living cell, at least within the near future; in fact, can we be sure that it is possible at all?

As we mentioned in the preceding section, the existence of a distinct difference between "living" and "dead" matter, was once generally accepted, and in fact it still is by some. According to this view, a cell could not be explained solely as a combination of molecules, but rather something new—called, among other things, the "life force"—must be assumed. A cell, then, was not subject entirely to the physical (or chemical) laws; instead, the "life force" was the true governing element that would cause these laws to work in a certain "purposeful" way.

It cannot, of course, be maintained that the cell is simply a cluster of molecules. It is obvious that something new appears when a cell is created from molecules. But it is equally obvious that an atom is not simply a sum of elementary particles. When these particles combine to form an atom, something new appears automatically, namely the combination of the particles, the interaction that gives the atom many new properties that the separate components do not have. To take a yet simpler example of the same principle: let us start with three straight lines of equal length, the straight line being one of geometry's simplest elements. By arranging them into an equilateral triangle, we come up with a number of new and unexpected properties: the sum of the angles is equal to two right angles, the height of the triangle is $\frac{1}{2}\sqrt{3} \times$ the length of one of the sides, and so forth. And if we proceed to combine several equilateral triangles of equal size, we can build up three and only three solid bodies: the tetrahedron, octahedron, and icosahedron, having 4, 8, and 20 sides respectively. If anyone doubts that new and unexpected properties result from the simple combination of simple elements, he is hereby challenged to calculate the relationship between the volumes of the octahedron and icosahedron!

The obvious fact that the cell has completely different reaction potentialities than do molecules (which are never as complex) should not be used, then, as an argument against the possibility that a cell might be built up entirely of molecules; rather this phenomenon is another, more complex example of a unique way of reacting that is entirely the product of the interaction between the constituent molecules. But yet it cannot be proved that a "life force" does not exist. It can be said only that we have no really compelling reason to assume, in principle, that anything new does exist in the cell. The more we study the simplest living units and the most complicated "dead" molecules, the less distinct the difference between them becomes. In fact, the cell, which is a very complicated structure, is not the simplest living unit we are aware of: for example, the chromosomes within the cell nucleus have a high degree of independence; the virus, once thought to be a very simple living being, is now generally considered to be nonliving, and it represents the intermediate stage between the two.

Before life appeared on earth, the different elements that make up the earth were combined to form different chemical substances. For the purpose of our discussion, the most interesting of these were the different carbon compounds. The air's carbonic acid was absorbed by the water, and, under the influence of the solar rays and the consequent temperature changes, more and more new combinations of carbon, hydrogen, oxygen, and possibly other substances were formed. In the course of millions of years, all of the possible different combinations were formed, most of them simple; but a few molecules of even more complicated substances appeared accidentally now and then. One of these accidentally acquired the property of being able to accept certain other molecules from its environment and to build with them a new molecule having its own structure. This molecule "multiplied" itself and dominated the small lake or pond in which this unobstrusive but important event had taken place.

Much more than this, however, was required for the appearance of "life." If a substance capable of multiplying itself became common, the chances became very great that one of its molecules would change accidentally (such a change could happen, for example, if several atoms exchanged places, or if a molecule incorporated some new atoms). As a result of most of these changes, the ability to multiply was lost. But in some of the new molecules, this property was conserved, and the result was a new substance having characteristics somewhat different from those of the first, but still capable of multiplying itself.

It was also possible for two or more such substances having different properties to join together to form a more complicated aggregate, which was capable of carrying out a greater number of complex chemical reactions. These were simple organisms.

Among these organisms a general competition soon prevailed. Those that could multiply the fastest and that had the most resistance to the different changes in their milieu had the greatest chance of survival. If, for example, the lake in which they lived dried out accidentally or froze, all of the organisms unable to endure this ordeal were destroyed, and only the hardier ones lived on.

The accidental changes that took place among the molecules continued among these extremely simple organisms—biologists call these changes mutations. Again, the result of most mutations is a diminished fitness for survival, and since this means the eventual, if not rapid, disappearance of that species, such mutations are of little interest to us. Much more important are those which produce species better able to survive, by increasing their ability to reproduce or their ability to resist external dangers. Many organisms increased their capacity for survival by becoming more complicated, and in this way increasingly complicated beings came into existence.

CELLS BUILD PLANTS AND ANIMALS

The cell is the smallest unitary constituent of plants as well as of animals, and a single-celled organism like the amoeba can therefore be considered a typical very primitive living being. The amoeba is by no means, however, the most primitive form of life we know. On the contrary, it is in a way the end result of a long development—that from the giant molecules to the first living beings. The further development from the amoeba to man occurred on another level, beginning when several cells joined together. It is from a study of the rich and varied world of giant molecules and primitive microorganisms that we may hope to gain a detailed description of the single-celled being. We have good reason to suspect that the step from the molecule to the amoeba is as long as that from the amoeba to man, or perhaps even longer.

A cell, as a rule, consists of a cell nucleus surrounded by protoplasm, which in turn is enclosed by a cell wall. The cell nucleus is the carrier of the cell's most important characteristics, and it regulates the cell's division, this being the way a cell reproduces. The nucleus contains a number of long striated threads, called *chromosomes,* which contain long spiral molecules. The chromosomes are the bearers of the *genes,* the heredity factors (or units of hereditary material) that principally determine the cell's reactions. When, through successive divisions, a fertilized egg cell develops into a multicellular being—a plant or an animal—it is the genes that determine the characteristics of this organism. It has been demonstrated that each heredity factor either is bound to a single giant molecule or actually constitutes one. If any one of these molecules changes, a mutation occurs, thus changing one or more properties of the being into which the egg cell is developing. Biological experiments can produce such mutations, one example being the exposure of a cell to a large dose of radiation; but these mutations occur also in nature, in which many are produced by weak cosmic radiation and by the radioactive rays that are found everywhere.

Currently we accept the general idea that biological development can be explained by mutations in combination with natural selection. In its essential parts, therefore, Darwin's theory of development has been accepted. In Darwin's time mutations were not known about; their discovery has led to extensive modifications of his theory, but it has also eliminated the most important objections to it.

On the whole, the development from the amoeba to man took place according to the same schema as the development from giant molecules to cells. If a plant or an animal mutates, one or more of its hereditary properties are changed in the process. As a result, the mutated individual and its offspring, like the mutated molecules, have either increased or decreased their proficiency in the struggle for existence. If the species is less fit for survival, it soon dies out, and the mutation has, in the end, been insignificant. But if the new species or variety has acquired the ability to feed itself more easily, to protect itself more ably from its enemies, or to multiply itself more quickly, it gradually surpasses those groups that have not improved upon their characteristics through mutation. Through a series of favorable mutations, and the selection process produced by the competition for existence, a continuous changing takes place. At the same time, a differentiation is constantly occurring. A certain mutation can, for example, enable a plant to survive a cold climate, and another can permit it to thrive in warmer environs. This species, then, will accordingly divide into two species, a northern and a southern, which do not compete with each other. It is also possible that both species might thrive side by side. The great numbers of species included in the plant and animal kingdoms show how many different species can exist within the same area.

Mutations and natural selection are the most important factors in the appearance of new species and the changes they undergo. There is also a third factor—bisexual reproduction —that makes the development proceed at a rather quick tempo. Assume that a change from one species to another requires mutations affecting a number of different genes. If the reproduction is unisexual in nature, so that each daughter individual inherits its mother's characteristics, all these mutations must take place throughout this very direct line of descent. If the reproduction is bisexual, each individual inherits characteristics from both its father and its mother, and it is possible for that individual to inherit all the mutations that appeared among its many ancestors. Bisexual reproduction is, therefore, necessary for the very great changes that are required for the appearance of the more complicated species. That we have managed to become human beings by this time is due to the fact that man and woman have again and again accomplished a mixing of their biologically best hereditary characteristics—and also their worst, but this has not been very important in the long run.

When we compare an amoeba with a human being—to confine ourselves to this particular segment of the long chain of complications—our first impression is that they are so completely different that they cannot be related in any way. The first and simplest explanation of the appearance of the different species was that a God had created all beings according to his own will, but this could not satisfy indefinitely man's instinct to investigate and order the natural phenomena. The more he studied the intermediate links in the chain from amoeba to man, the more convinced he became not only that the chain is very long, but also that each link interlocks with the one preceding it and the one following it. There are still many "missing links," but in all probability these are due to our imperfect knowledge. Today almost no one doubts seriously that the single-celled animal was developed on a long and complicated path to multicellular and increasingly complicated beings, among them man.

Once we realize, then, that the amoeba has developed into something as wonderfully and ingeniously constructed as man, our reaction is that some special force must have led the development in precisely this direction. Was this force God or perhaps just nature's harmonious progress toward higher forms? Perhaps nature works according to the principle, *Warum es einfach machen, wenn es auch kompliziert geht* (Why make it simple, when it could be made complicated as well)?

The simplest and most sympathetic answer is naturally that God or some other "force" directed the development toward a certain goal, and our self-conceit makes it easy for us to imagine that this goal is "the crown of creation," that is, ourselves. Viewed on a large scale, with the amoeba as the starting point and man as the end point, this perspective appears most natural; but as soon as we begin to look closely at the details, we see that this conception is untenable. The question is this: How does this supernatural intervention take place? For if the natural laws are, without exception, valid, there is no place left for such an intervention.

It is true that the entire enormous research field of biological development is a long way from being so thoroughly investigated that we can reconstruct with certainty every detail of this development. But we are beginning to understand which are the most important factors. We are beginning to see that the awesome wonder of the evolution from amoeba to man—for it is without a doubt an awesome wonder—was not the result of a mighty word from a creator, but of a combination of small, apparently insignificant processes. The structural change occurring in a molecule within a chromosome, the result of a struggle over food between two animals, the reproduction and feeding of young—such are the simple elements that together, in the course of millions of years, created the great wonder. This is nothing separate from ordinary life. The wonder is in our everyday world, if only we have the ability to see it.

We know too that it is incorrect to say that the amoeba has developed into the human being. As we observed in the preceding section, the contemporary amoeba is the result of a process as long as that which has engendered human beings. Both derived from a primitive cell. One series of mutations, then, after hundreds of millions of years, produced man; another series, the contemporary amoeba. The latter process may not even have been one of increasing complexity, but rather one of simplification. The decisive criterion for determining whether a mutation—or a series of mutations—has yielded a biologically enduring result is whether it increases fitness for survival, and simplification can certainly increase that fitness. If, for example, a human being lost his appendix through a mutation, his construction would be simpler and at the same time better adapted, since the risk of dying from an inflammation of the appendix would have been eliminated.

The biggest difference between a change that produces complicated beings and one that does not is that the former draws more attention. Biological simplification occurs as often as complication, and it is only because people have found the complicated so much more interesting that they have come to believe that these changes should take some kind of precedence.

Not only, then, is there no reason to search for a special "force" that impels an ever increasing complication, but in fact if we consider the composition of the entire universe we see that the processes of complication are quite minor. Most nucleons are still in the form of free protons, hydrogen nuclei, and only a relatively small number combine to form heavier atomic nuclei. In the universe, free electrons outnumber those bound to atoms, and free atoms are more common than those bound in molecules. But on the earth itself, the majority of the electrons are bound in atoms; and most atoms, in molecules. This is due to the fact that under the special conditions (temperature, pressure, and others) that prevail on earth, these combinations are more stable. Transformations from the simple state to the complicated proceed more easily than the reverse. Whereas the free protons dominate in the universe, on the earth free molecules dominate, greatly outnumbering those that have combined to form cells. But of the cells, those that have combined into more complicated organisms outnumber the free, or single, cells. These proportions are the result of the competition between building, complicating processes and those which break down and simplify.

THE FORMATION OF SOCIETIES

The last step in the long chain of complications is the combination of plants and animals to form societies. Most striking is the formation of societies among animals, especially man; but in the plant world as well, similar relationships are very important.

In the plant world it often happens that one plant lives off another, either partly or wholly: a parasite draws nourishment from its host-plant. Also two may live in a state of *symbiosis,* extracting nourishment from each other, or helping one another in some way. A common example of symbiosis is the bacteria that live on the roots of pea plants and form bumps there. These bacteria get their nourishment from the pea plant; but since they are able to take nitrogen directly from the air, and the plant cannot, they help the pea plant by supplying it with nitrogen compounds. From this simple example of symbiotic cooperation between two different plants, we can proceed to more complicated types of cooperation. The science that treats the coexistence of plant forms is called *ecology,* and the more it is developed, the more clearly it shows us how enormously complicated is the interaction between different plants of a forest or a desert. If it is to thrive, a plant requires not only a favorable climate and good earth, but also good neighbors, which to a certain degree do compete, but which also help each other significantly.

Within the animal world the types of units formed for the purpose of mutual help are extremely varied. The smallest unit is the family. The mother, if not both parents, gives the young ones food and shelter, which is essential to the continuation of the species. Sometimes larger units are advantageous. A pack of wolves unite to hunt more effectively; boars or elephants join together in herds for mutual protection against enemies. Yet wild cats and lions find it more to their advantage to hunt alone; and the hare's defense against enemies is such that numbers of hares are no more effective than a single one. Perhaps at one time in the course of evolution there did appear a member of the cat family whose tendency was to hunt in packs; but since this method produced less food per animal and no advantages, this type soon died out.

For some animals it has proven advantageous to form large, well-organized societies. The most distinctly social formations are those of ants, bees, and man.

At first there would seem to be an enormous difference between the cellular coalitions discussed in the preceding section and the coalitions of animals into societies, because cells grow into a unit, whereas the people or bees or ants of a society have freedom of movement. But if we reflect on the long chain of complications that we have traced thus far, from the simplest units we know to increasingly complicated formations, the social arrangement is a natural last step.

In fact, many analogies between the men in a society and the cells of a body can be drawn. Among the cells there is a continual exchange of nutrients analogous to the exchange of goods among men. The body's nerve cells, which quickly send messages and orders to the different parts of the body, correspond to the telegraph and the telephone. The cellular specialization in different functions within the brain, stomach, or muscles corresponds to the division of labor in a society that yields teachers, farmers, or factory workers.

It should be emphasized, however, that it is invalid—and could in fact be disastrous—to draw certain conclusions from an analogy. Thus, although we generally consider man to be a "higher" being than the cells he consists of, we need not conclude that the State is on the right side in the perpetual struggle between the individual and society. Actually, if we weigh these two adversaries against one another, we observe that a man, when stripped utterly of society, is not a man. The slogan "Let us go back to nature" is a salutary one in the fight against the increasing preponderance of society; but the fact that few follow it is a sign of how great the advantages offered by society actually are.

The question whether democracy or dictatorship is the most stable social form cannot, of course, be answered by an analogy with the human body. For example, it might appear as if the body is ruled by a strict dictatorship, because it is true that the brain can arbitrarily dispatch commands to the different parts of the body. But in fact there are severe limitations to the

The Long Chain of Complications

brain's power, since the body's parts have effective means of opposing its commands. If the brain has directed the legs to walk a number of miles, the leg muscles send out an increasingly strong protest in the form of fatigue and pain, until the protest becomes full revolt: the legs give in to the fatigue and refuse to submit to the dictatorship any longer. Finally the general revolution occurs: the brain's rule is suspended until its possessor has slept away his weariness.

The great advantage the central nervous system has given the animal is the ability to react more quickly, a factor of decisive importance in the struggle for survival. The necessity for attack and defense was at least one of the reasons for the evolution of such a nervous system. Similarly, the same needs gave rise to the development of human society. A strong governing power is necessary in all states, which must often fight for their existence. The stronger the governing force, the greater the temporary advantages it brings with it; but since voluntary cooperation without state compulsion is more effective in the course of a long period of time, the advantages of dictatorship are indeed only temporary. The needs of a society, then, might be said to be a balance between the long-term requirements and the short-term, or urgent, needs; and to some degree this balance is one of the reasons for the fluctuations between democracy and totalitarianism.

To predict, on this basis, whether a democracy or a dictatorship will be the final form society takes is naturally impossible. What is of greatest importance is that biological development has now reached a unique state. One might expect that before long—at least within a couple of generations—the entire world will have been united into a single state, quite simply because the present division into separate states is unstable since it leads to war. In such a united state, there would be no real need for a dictatorship because there would be no competing state against which to struggle. But because of the advent of society, for the first time in the course of biological development the necessity for the individual organism to have maximal efficiency has diminished. In such an atmosphere, a dictator capable of any evil can persist, as long as the state is internally stable; unfortunately the art of producing such stability is within the ability of a dictator.

III

Atoms and Men

THE SENSE ORGANS AS PHYSICAL INSTRUMENTS

We have traced the long chain that has led from the elementary particle and the atom to beings as complicated as man. It might now be asked whether this development has removed us from the realm of the atom or whether atomic phenomena still have a direct influence upon the human mechanism. We may answer this question by studying the functioning of our sense organs.

If a physicist wishes to measure the intensity of a beam of light, he often uses a photocell, an instrument that when illuminated produces an electrical current that is proportional to the intensity of the light. If the light is too weak to produce a current measurable by an ordinary photocell, this instrument can be coupled to an amplifier of approximately the same type that amplifies radio signals in an ordinary radio set. With a photocell that has been coupled to an amplifier, it is possible to detect even very weak small intensities of light. We can proceed even farther if we replace the photocell with a photomultiplier or other light-sensitive tube. With these devices the sensitivity can be increased to the point at which single photons can be detected.

As we observed at the beginning of the second chapter, light consists—according to the particle theory—of a stream of photons. The smallest quantity of light that it is theoretically possible to detect is a photon, because smaller quantities of light simply do not exist. A photon cannot be directly detected. Its existence is revealed in other ways; for example, it may collide with an electron, which is thereby set in motion that is detectable in several ways. The photons of X-rays and gamma radiation are so rich in energy that the electrons with which they collide acquire high velocities and are relatively easy to detect. With the most sensitive instruments we have for measuring gamma radiation, we can obtain a signal for almost every photon that reaches the measuring device. Therefore it is possible to count directly the number of photons that arrive, and with this achievement we have attained the limit of what is theoretically possible. But the photons of visible light have a much lower energy content, and therefore their ability to dislodge electrons that can be detected is reduced. Using, again, the best instruments yet constructed, we can detect no more than an average of every fifth or tenth photon.

How does the human eye compare with one of these highly sensitive instruments? The eye's sensitivity varies greatly with external conditions. In strong sunlight the sensitivity is reduced, and it is at its maximum only when the eye has had some time to adapt itself to very weak illumination. The degree of sensitivity also varies with the color, that to yellow being greatest and that to red and to blue being less. In order to measure the eye's sensitivity under the most favorable conditions possible, we measure the sensitivity to yellow light after the eye has been adapted to darkness. This kind of investigation has shown that the weakest light signal the eye can detect corresponds to the few (five to ten) photons which enter through the pupil and strike the retina. The eye therefore, under the most favorable conditions, exhibits the maximum sensitivity that is physically possible.

The retina of the eye contains a light-sensitive substance (visual purple), which is transformed when light strikes it. This transformed substance causes a neural stimulation, which is conveyed to the brain, where it is perceived as light. If several photons enter through the pupil, no more than one or two can be expected to strike the light-sensitive substance, and a photon changes no more than a single molecule of visual purple. Thus the smallest quantity of light that can be perceived by the eye corresponds to the transformation of a single molecule—or possibly two or three.

The function of the eye, then, is directly dependent upon atomic phenomena: a single quantum of light is detectable—obviously under the most favorable conditions—by the human eye.

The sensitivity of the ear, too, is the lowest of what is physically possible. If a physicist wishes to measure a sound, he uses a microphone in combination with an amplifier. When sound waves strike the microphone's membrane, this causes vibrations, which produce electrical currents that can, in turn, be amplified. The more sensitive the microphone and the amplifier, the weaker the sound that it is possible to detect. The limit that can be reached is determined by heat motion. As soon as a body has a temperature higher than -273 degrees Centigrade, known as "absolute zero," its molecules are in perpetual motion. The membrane of the microphone is struck by the air molecules and is thereby set into motion. In addition, the molecules that the membrane itself consists of also move in an irregular manner and thereby contribute additional irregular motion. The vibrations are extraordinarily small, but the measuring techniques are so highly advanced that even these vibrations are detectable. If a sound is so weak that the vibration it produces in the microphone's membrane is smaller than the vibration caused by the heat motion, it is of course impossible to establish its existence as a physical fact. We cannot properly define, then, a vibration as a "sound," unless it is stronger than the heat motion. With a sensitive microphone and amplifier it is possible, therefore, to detect any sound whose vibration is several times greater than the heat motion.

The ear's sound-sensitive organ is located within the inner ear, and when a sound strikes the ear-drum membrane, it is conducted into the inner ear. In this way the sound-sensitive organs are made to vibrate, and this is perceived as sound. The sensitivity of the ear varies with the pitch. It is greatest if a tone has several hundred vibrations per second, corresponding approximately to the tenor octave. The minimum strength that a sound of this optimal pitch can have, and still be perceived by the ear, corresponds to a point at which the vibrations in the inner ear are several times greater than the heat motion in the inner ear. The sound-sensitive organs in the inner ear, like all other objects, are affected by heat motion and therefore undergo constant small vibrations. If this "natural" state is disturbed by a sound striking the ear, the sound's vibration need be only several times greater than the heat vibration in order for the ear's sound-sensitive organs to react. A sound, then, can be perceived as such if, once it is within the inner ear, it meets the physical definition of "sound" specified above. It is true that the outer ear could have been constructed so that it picked up more sound, and that the passage into the inner ear could have been made more effective. Also, it might have been a further advantage if the ear were sensitive to both high and low pitches. But these hypothetical observations notwithstanding, the inner ear's sound-sensitive organs, on the whole, are as sensitive to tones within the optimal pitch interval as any physical instrument designed for this purpose. The ear functions within the limits that have been set by the atomic phenomena involved.

The sense of smell differs from the other two in that it is more difficult to evaluate its sensitivity in terms of the same kinds of precise atomic phenomena. The theoretically smallest quantity of a substance is a molecule; but for a smell to be detected, it is necessary that a large number of molecules reach the mucous membrane of the nose. This might lead us to say, therefore, that all people—like the old fisherman of the traditional Swedish tale—see well, hear well, but smell terribly. But such a statement is probably misleading, since the quantity of a substance required to give the nose the sensation of smelling is often much less than what a chemist could detect.

The human body's direct relationship with the atomic world, as exemplified by the functioning of the eye and the ear, can perhaps be said to be a valuable and well-administered inheritance from our very oldest ancestors, the organisms that consisted of only a few molecules and were as different from the amoebae as these are from us.

THE NERVOUS SYSTEM AND THE ELECTRICAL IMPULSE TECHNIQUE

The increasing complexity of the nervous system has been of decisive importance for the development of the higher animals, and especially of the human organism. It is not because of a superior bodily strength, a greater agility, or a prodigious capacity for reproduction that man has become the master of the earth; rather his supremacy is due to the fact that his nervous system has become superior to that of any other animal. The immense organization of nerve cells in his skull has made him cunning, contemplative, and systematic, and in the struggle for world hegemony these characteristics have been more crucial than the elephant's strength or the tiger's agility; recently he has even succeeded in restraining the enormous reproductive powers of insects and bacteria, thanks to his orderly scientific mode of thought.

The great significance of the nervous system is its function of coordinating the reactions of the different parts of the body. The unicellular animal's single cell is directly affected by the outer world. Its ability to locate nutriment or avoid a poison is a consequence of the direct effects of these substances upon that cell. In a multicellular animal only those cells on its surface can be directly affected from the outside; but if the entire animal is to react optimally, it is necessary that, in one way or another, the stimuli from outside the organism be communicated to all of the cells. Some of this communication is chemical, for cells affect each other by a continuous mutual exchange of substances. But chemical contact between the cells takes place very slowly, especially in higher animals, because it takes time for a chemical substance to diffuse through the organism. To meet the need for more immediate intercellular communication, certain cells have evolved in such a way that they are long, threadlike, and very excitable— these are the nerve cells. A stimulation of one end of a nerve cell causes almost immediately a disturbance throughout the entire cell. If a nerve fiber connects two of the body's organs in this way, the state of one can be reported quickly to the other.

If one end of a nerve fiber is stimulated, the normal chemical state around the point of stimulation is disturbed, and a system of electrical currents is produced in the area of the disturbance; these currents in turn disturb the chemical state farther along the fiber, where new currents are then produced. In this way an electrochemical impulse arises that reproduces itself along the nerve fiber at a velocity that, among warm-blooded animals, is as much as 50 meters per second. When the impulse has reached the other end of the nerve fiber it can stimulate a muscle, causing it to contract. (It is this very mechanism that causes a person to start upon receiving an electric shock.) The stimulation of another nerve can cause a gland to increase its secretions: for example, saliva and gastric juices are secreted when certain nerves signal the presence of hunger. Perhaps most interesting of all is the fact that an impulse in a nerve can be conveyed to one or more other nerve cells. This takes place in what are called the *synapses,* points at which a number of different nerve cells are joined together. A stimulation coming from one part of the body can therefore be conveyed, via the extremely complicated nervous system, to many different organs simultaneously and produce in them an often very complicated system of reactions.

Every impulse that is sent out when a nerve cell is stimulated has a certain strength and duration independent of the strength and duration of the stimulus. But of course a stimulus can be very weak, very strong, or somewhere in between, and what this degree of strength does affect is the impulse *frequency,* that is, the number of impulses that are sent through the nerve each second. For example, a nerve from a sensitive point in the skin does not normally send impulses to the large receiving centers in the spinal cord and the brain. Upon receiving a weak stimulus from that part of the skin, the nerve sends infrequent impulses as long as the stimulus persists. The brain perceives these signals as a weak sensation of pain. If the stimulus is increased, the number of impulses per second increases, but each impulse has the same magnitude. The more frequently the impulses arrive in the brain, the stronger is the perception of pain.

We cannot help but notice the similarity between the nerves' electrical telegraph system, which permits the rapid communication between different parts of the body, and our utilization of electrical signals, which is binding humanity together more and more closely into a huge social organism that reacts quickly and nervously. Interestingly, electrical technology often uses what is in principle the same type of mechanism present in the nervous system: communications of different kinds are today sent out in the form of a series of identical impulses, and the "information," that which is to be communicated, is determined by the lengths of the intervals between the successive impulses.

This similarity, however, is perhaps more accidental than we might at first realize. Electrical technology has not consciously imitated the nervous system's method of communication, but it has been forced to adopt a method that resembles it by the severe demands of technical efficiency. There is, of course, also a big difference between the conditions necessary for physiological communications and those required for electrical communications, which take place across great distances. The most important difference is that the nerve impulses consist of electrochemical disturbances that are propagated at a speed of scarcely 50 meters per second, whereas the impulses employed in electrical technology are electromagnetic and can be sent at the velocity of light (186,000 miles per second). Consequently, a communications technician can send out as much information as can thousands or even millions of nerve cells. Furthermore, a nerve cell can communicate impulses of only one magnitude; but in communications technology it is possible to convey information by varying the strength of the signals. Since this method is technically very simple, it was the first one to be employed and it is still the most common. In an ordinary telephone the microphone transforms the sound waves into electrical currents which vary according to the sound waves. It is therefore a direct "electrical picture" of the sound that is sent out along the telephone wire and is changed back into sound at the other end. But for telephony across long distances, a method resembling that of the nerves has proven advantageous: before transmission the electrical picture of the sound waves is transformed into a series of identical impulses in accord with quite a complicated coding system. When the "coded" communication reaches its destination it is "decoded," that is to say, it is changed into alternating current and then into sound.

VISION AND TELEVISION

Now let us compare the technological and the physiological methods of transmitting a picture. Using the television set as our example of electrical technology, we shall study the conditions required for transmitting a picture in the form of signals in a cable, and then we shall ask ourselves if these might be applied toward our understanding of the functioning of the optic nerves, the cable that connects the eye with the brain.

In a television set a picture to be transmitted is divided into perhaps 100,000 points and then information about each of these points is telegraphed to the receiver: that is, whether the point is light or dark, or (if it is color television) the color of a point. The entire telegram, containing the data on the light strength or color for the 100,000 points, takes no longer than one-twentieth of a second. If television is to reproduce movements that impress the viewer as being natural, it is necessary to send about 20 pictures per second.

The most interesting technical problem is the incredible speed of the telegraphy that is necessary. Within one-twentieth of a second 100,000 telegrams, one for each of the picture's constituent points, must be sent. Since the picture becomes sharper as the number of points into which it is divided increases, perhaps television's most important objective is to send as many telegrams as possible per second—or, in the current idiom, as much "information" as possible. For this reason, extremely careful analyses have been made to determine what factors limit the amount of information that can be sent via an electrical current having certain properties.

The results of this analysis can be applied to the nerve fibers. We know that the maximum amount of information that can be sent along a nerve fiber is much less than what can be sent along an electric current-carrying wire. This is quite natural since the speed of the nerve impulses is so much slower than that of the electromagnetic signals. Since it is often important to send much more information than that which can be conveyed by a single nerve fiber, the only possibility is to send the information along many parallel fibers. This in fact is the way in which the problem is solved: many of the nerves contain thousands of fibers, and even if each separate fiber cannot convey a great deal of information the total capacity of all the fibers in the nerve is very large.

Once we know the properties of the nerve fiber and the number of nerve fibers in the optical nerves, it is possible to calculate the maximum amount of information that can be sent from the eye to the brain. We find that this amount is still much less than that required to convey a television picture. This means that the picture we see cannot be telegraphed in its entirety from the eye to the brain. In other words, the optical nerves cannot supply the brain with information about how every part of our visual field looks at every second. The telegram from the eye to the brain can therefore report only certain essential features of what we see—features which, for one reason or another, are of interest. The eye, then, is not a "camera," which perceives a picture automatically and then sends it on to the brain for interpretation. Rather, what the eye perceives undergoes an initial interpretation and systematization in the nerve centers that adjoin the retina. The final interpretation that is sent to the brain is a kind of "coded" communication, containing certain essential features of the visual picture in a format that is sufficiently concentrated to enable the optical nerves to effect the transmission.

We know little about either the principles governing this interpretation or the form in which the coded communication is relayed to the brain. It is possible that certain conclusions can be drawn from the fact that entirely different pictures can give similar impressions. By drawing no more than a few black lines on a piece of white paper, an artist can produce a sketch that is immediately recognizable as a likeness of the model from which he has made the sketch. It is conceivable that this recognition is due to the fact that the synthesis made by the optic nerve centers when the eye focuses on the model is very similar to that made when the eye focuses on the charcoal drawing; if this is true, then the signals received by the brain for both objects have a certain similarity. Viewed objectively, a color photograph must convey a sensory impression very much like the one received directly from the model, whereas a drawing can consist of a number of black contours that in reality do not exist. And yet a drawing can seem more "real" than the photo! A skillful artist understands intuitively the rules by which the eye's nerve cells code their communications, and his art consists in making the viewer's optic nerve cells send telegrams to the brain that contain more truth than those elicited by a photograph.

MATHEMATICS AND MACHINES

One of the most difficult problems treated by philosophers through the ages is the relationship between the mind and the body: What kind of relationship connects the spiritual life with the brain? We have no reliable reason for assuming the existence of a "soul" separate or independent from the matter of the brain; but does this imply, then, that the soul is simply a function of the brain? It has often been maintained that thoughts are no more than "chemical reactions" in the brain, but many people protest vehemently against such assertions. The intuition in man still argues that no matter how fine the gray substance within our skulls, it is coarse in comparison to "such stuff as dreams are made on."

We shall not enter into a deeper analysis of this extremely intricate problem, but it might be of interest to discuss, as an analogy, a similar but incomparably simpler problem, namely, the relationship between the material parts of a mathematical machine and the mathematical operations it can carry out.

First let us examine a very simple small machine, the desk calculator. It contains nine cogwheels, each wheel having ten cogs, and each cog representing one unit. If the first wheel is turned forward six cogs the number six is denoted. The second cogwheel represents the tens, a third the hundreds, and so forth. With these nine wheels we can represent all the numbers that can be written in nine digits, that is, all numbers up to 999,999,999. If the number 0 is included in the calculation these nine cogwheels can denote one billion different numbers. This total of ninety different cogs can, in other words, be combined in a billion different ways.

If the calculator has a second set of nine cogwheels, all of the numbers up to one billion can, of course, be represented by it too. If it has several more cogwheels, it can add or subtract the figures represented by the first two sets of wheels. Still more complicated machines can multiply or divide the figures.

Mathematics is regarded by many as the most abstract of human mental processes. Many thinkers, from the ancient philosophers to Jeans, have refined their concept of mathematics to the point at which they have chosen to think of God as a mathematician; the Pythagorean metaphysics takes the integer as its point of departure. How is it then possible that something as abstract, or as "spiritual," as a number can exist in a machine consisting only of such coarse materials as a bunch of oily cogwheels? How is it possible to effect certain mathematical operations like addition or multiplication by the mere rotating and meshing of these wheels? If there are initially two numbers in the machine, addition produces from them a new number which was not originally there. More complicated machines can solve problems of such complexity that, without the machine's help, they would simply exceed our ability to solve them.

Is it not apparent—one may ask—that the machine consists of two dissimilar parts, one material, and the other immaterial and mathematical? An engineer whose task it is to repair the machine must concern himself solely with its construction; but a mathematician who is delving into the beautiful mystery of numbers will deal only with the immaterial part. For him the rattling of the cogwheels during the carrying out of a calculation is, if anything, an irritating distraction. In time he perhaps begins to believe that the nonmaterial component is not only much finer than the material one, but entirely independent of it. His suspicion is substantiated when he finds that the same mathematical operations can be carried out in machines of a completely different construction. The cogwheel is not even necessary! A calculating machine can consist, instead, of electrical relays or transistors; or brain cells. This exemplifies the degree to which the machine's nonmaterial element is independent of the material one.

It is possible to probe more deeply into the metaphysics of the calculating machine. Assuming, then, that the mathematical operations are parallel with, but independent of, the movements of the cogwheels, some might wish to ascribe a soul—though naturally a small one—to the calculating machine. But no one would maintain that the nonmaterial element of the machine guides the material element, for this would mean that the machine solved a problem by a purely "intellectual" effort, and then by an "act of will" rotated its cogwheels to indicate the result. It seems much more reasonable to say that the "nonmaterial" part of the calculating machine consists of *combinations* of the positions of the material elements. The nine ten-toothed cogwheels are more than merely nine ten-toothed cogwheels: together they are a billion combinations of positions. It is these combinations that represent the numbers, and the calculating operations occur as the combinations are changed according to certain rules. The combinations are independent of their material basis in the sense that the same combinations can be produced by small cogwheels or large, or by transistors instead of cogwheels. We can work with the combinations perfectly well without taking into account what is actually combined. If we wish, we can refer to the combinations as the "spiritual" element, but this is still introduced in a rational way as an inevitable consequence of the problem being analyzed.

The more elements with which we start, the more combinations that are possible. This applies not only to calculating machines, but also to atoms and everything that can be built up from them. In the previous chapter we followed the long chain of complications, the sweeping growth of combinations from the atom to the human being: is it possible that it is the enormous number of possible combinations in his brain cells that has given man his soul?

THE COMPUTER

The technology of electrical impulses has assumed epic significance as a result of the rapid development of the computers, which are in the process of revolutionizing society. This dramatic achievement is the result of the increasing demand in many fields for complicated computations that can be carried out quickly as well as accurately. The computers differ from the ordinary, simple calculating machines in two respects: their speed is fantastically greater, and they are capable of directing the course of their own computations according to a preassigned plan, or program. The construction of computing machinery has been made possible by electrical technology. The first computers denoted a number by sending out a number of electrical impulses along different circuits. Like the different cogwheels in an old calculating machine, one circuit denoted the units, another circuit, the tens, a third circuit, the hundreds, and so forth. But most of the computers in use today express a number by means of a binary number system; this means that the number can be represented by a series of impulses in a single circuit. The numbers with which the computation begins are first introduced into a particular part of the machine called the *memory*. Here the number is stored in the form of a magnetized tape or a system of electrical currents. The memory has the characteristic of being able at any time to send out the impulses representing the number stored in it. If, for example, we wish to add two of the numbers contained in the memory, both of these numbers are sent to an addition unit. When two impulses enter it at the same time, the addition unit sends out a series of impulses which correspond to the sum of the two numbers it has received. This sum can then be sent back to another part of the memory and preserved for possible use later on in the computation. Subtraction, too, can be done in the addition units. A multiplication unit can multiply any two of the memory's numbers.

The multiplication of two ten-place numbers is the kind of tiresome job one prefers to avoid. A computer provides the answer in less than one-hundredth of a second. The computer can calculate much faster than the human brain because the electrical impulses in the machine's circuitry can be propagated so much faster than the impulses in the brain cells that course through complicated paths when a person carries out a computation in his head.

So the computational speed of the human brain is greatly inferior to that of the computer. But in its ability to judge and evaluate the result of a computation, the brain is still superior to any machine yet invented. If a mathematician wishes to make a complicated calculation and has an assistant to help him, he can instruct the assistant in the following way: first make this calculation; if the result is larger than a certain figure, proceed with a second calculation; but if the result is less than the given figure, perform instead a third calculation. In other words, the computational assistant calculates according to a certain plan, but this plan necessarily changes in the course of the computation according to the results obtained.

It has proven possible to build computing machines that can direct at least the details of their own computations. Besides the addition units, the multiplication units, and the memory, this type of machine has a *control unit*. This unit initiates the different calculations and sends out the impulses that determine when a number in the memory is to be sent out and whether it will be sent to the adder or to the multiplier. In other words, the control unit causes the computations to be performed in the right order according to the given plan. But it can also change this plan in the course of the calculation in accordance with a given program. To take a simple example, we will try to find the square root of 10, that is, the number which, when multiplied by itself, gives 10. The machine can be instructed, or "programmed," to feel its way forward: that is, having multiplied the first number by itself, it will try a larger number if the result is less than 10; but if the result is greater than 10, it will try a somewhat smaller number. The machine begins by demonstrating to itself that 1×1, 2×2, and 3×3 are less than 10, but that 4×4 is greater than 10. The machine then tries 3.1×3.1, and finds that it is too small, but that 3.2×3.2 is too great. It then tries 3.11, 3.12, and so forth. Within a fraction of a second it has attained the correct answer, 3.162!

More complicated problems, for which the machine must carry out millions of computations before reaching the final result, can take as much as hours, or even days, of machine time. The machine, then, treats the numbers in its memory according to the program of instructions in its control unit. In some types of machines, the entire process can occur without any movement within the machine. But although it is quiet, it is "thinking" intensively about the problem by sending impulses between its different organs; but even the paths that these impulses follow are regulated by electrical currents in transistors or electron tubes, which in turn are regulated by other impulses. The computer of today cannot replace a mathematician, since only he can formulate a problem and then interpret the results of the computations; but it *can* replace a well-qualified assistant—in fact, from the standpoint of speed, it can replace a whole staff of such arithmeticians. The mathematician instructs the assistant by speaking to him, and the machine by pressing certain buttons. In this way begins the computation in a human brain or in a machine. We have a detailed knowledge of the paths taken by the impulses within a machine; but our understanding of the ways in which human nerve impulses course through the different parts of the brain is still limited. The more computers that are built, the better will be our comprehension of the solution of mathematical problems by electrical technology; and since the mathematical operations that take place in the brain may in certain respects be similar to those of electrical technology, perhaps we shall in this way increase our knowledge of the processes of human thought, or at least certain aspects of them.

The human mental operations that can be carried out with the use of a computer are not only mathematical. Rather, any mental activity that occurs in conformity with a given plan can be performed by such a machine, even if the plan is very complicated. For example, it has proven possible to program a computer to play chess or checkers. The machine is informed of the opponent's moves and determines which moves it can make according to the rules of the game. Then it calculates the moves with which the opponent can respond and those with which it itself can respond in turn, and so forth, until it selects the possibility that, according to certain rules, it finds to be the most favorable. Then it moves its pieces accordingly. How skillful the machine is depends on how complicated it is. A machine able to compete with one of the great chess masters would be enormous and very expensive; but a computer of ordinary size can be programmed to play correctly, and it can even beat a beginner. Computers are reported to be extremely clever in checkers.

It can be contended, of course, that the similarity between a calculating or chess-playing automaton and a human being who calculates or plays chess is no more than a superficial analogy. We do not know enough about the physiology of thought to resolve this argument with any certainty, but as we have already suggested, there is much to indicate that the similarity is not accidental. Both the nervous system and the machine make use of electrical impulses that can be coupled in different ways. In the machine, the relays, vacuum tubes, or transistors choose between the different paths. In the nervous system, the synapses coupling points between different nerve fibers have a similar function. Although the basic circuits are different, then, the constructions would appear to be similar. It can be further objected, certainly, that the achievements of the machine—the mathematical calculations and perhaps also the game playing—cannot equal the most refined expressions of human abstract thought. But, on the other hand, it is apparent that the construction of the brain is incomparably more complicated than any computer in existence: for such a machine has perhaps 10,000 transistors, which can carry out different couplings; but the number of synapses in the brain that can direct nerve impulses through different paths is in the billions. On the basis of our current knowledge of computers, we can thus assume that, should an apparatus contain as many coupling elements as the human brain, it too would be capable of reactions as complicated as those that take place in the brain.

A NEW LINK IN THE LONG CHAIN OF COMPLICATIONS?

Is it possible that, with the advent of the computers, a new link has appeared in the long chain of complications? In the preceding chapter, the chain had ended with the human being. Now we might ask whether man will indeed be the last step in this development for all time, or whether, as Nietzsche fantasized, he will defeat himself and make way for an *Übermensch*. From a less poetic and more scientific perspective, the question is whether a more complex aggregate than the human being—or, in other words, a system having more possibilities for combination—can arise.

Although the computers of today can perform mathematical and logical operations that are beyond human capacity, it is true that their complexity does not approach that of the brain. But it must be remembered that the computers have existed for only a few decades and have developed at a startling rate. In more and more areas they are becoming capable of replacing human beings: they can coordinate the entire body of international air traffic; they can manage the accounting and bookkeeping of a large business; and they can keep track of the economic factors involved in governing a nation.

Since these machines are capable of all of these functions within so short a period of development, we are now compelled to ask where computer technology will lead in the future. As a result of one significant new innovation, the workings of a modern computer are no longer necessarily limited by a set program. Computers can improve their own programs, learn from their own mistakes, and work out methods aimed at the goals that were initially provided. In this and a number of other ways the processes that a computer can carry out are approaching the working methods of the human mind, and it is not too much to predict that in an increasing number of areas the machines will prove superior to man. Already, certain new data techniques are being developed by the computers themselves. There is therefore some basis for the speculation that the computers will come to regulate an increasing number of areas in society and thus cause it to function more and more independently of the individual. One vision of the remote future is the possibility of a completely computer-controlled society: a virtual nightmare, in the opinion of many people. But then it may also turn out to be an ideal society.

To look again at the long chain of complications, recall that what we observed was a path from atoms and molecules through cells and all of the subsequent organisms to man. This new link is not a direct continuation of that chain: for although it is a new complication, it is not based upon protoplasm or cells, but upon metal wires, transistors, and other coupling elements. But the material basis of this new variety of complications is not, after all, of decisive importance. It is its "ideas"—that is, its combinations—that constitute the "new link's" potential effect upon human existence.

IV

The Cosmic Perspective

THE CREATION

If we contemplate existence itself, our thoughts can occasionally lead us both backward and forward in time: back to our own birth and perhaps also to the origin of the human race, and forward to our imminent death and to mankind's final destiny. And if we wish to penetrate yet further, we confront the awesome problems of the world's creation and its final extinction. In this chapter, we will discuss the creation.

In past years the traditional library in a Swedish home was said to consist of two books, a Bible and an Almanac, and these were considered sufficient for clear answers to both how and when the creation took place. The first part of the question was answered in the first chapter of the first book of Genesis, and the second by the Almanac. If we consult the Almanac for the answer to the second part, we find that, according to the Hebrew calendar, this is the year 5729 after the creation, and that it began on September 23, 1968 A.D.

An estimate of the world's age in terms of several millennia was characteristic of the ancient cultures of the Mediterranean region and the Near East. Thus the ancient Persians, for example, thought that the entire course of world history from the creation to the last judgment would have been run in 12,000 years. In other cultures, grander ideas evolved: according to one Chinese estimate the world's age would now be 129,600 years; and in the ancient Indic philosophy, we find lengths of time so enormous that even the time scale of modern natural science appears small. It was believed that four ages succeeded each other: gold, silver, bronze, and iron. This idea was prevalent in Western thought, too, as we know from such works as those of the ancient Greek and Roman poets. But whereas the Mediterranean philosophers estimated the length of an age at some thousands of years, the Indians calculated the Golden Age (Kritayuga) to be 1,728,000 years, the Silver Age (Thretayuga) to be 1,296,000 years, the Bronze Age (Dwaparayuga) to be 864,000 years, and the Iron Age (Kaliyuga) to be 432,000 years. Together the four ages formed a Mahayuga (literally, Great Age), having a respectable length of 4,320,000 years. The Indians carried this further. As soon as a Mahayuga was ended by an unhappy Iron Age, the world was cleansed and a new Golden Age began, and in this way world history proceeded until it had run through 1,000 Mahayugas, which constituted a Kalpa, or Brahma's day, with a length of more than four billion years. When Brahma began his day he created the world, and at the end of the day—after a thousand Golden Ages—he destroyed it, assuming the shape of Shiva, the divine destroyer. Brahma repeated this each new day, and when he had lived 100 Brahma-years he would have created and destroyed the world 36,500 times. Anyone who might have been disturbed about the state of the world at this remote point in time could comfort himself with the fact that when Brahma, by this time age-weary with good cause, died, he would be reborn in a rejuvenated and more powerful form.

The Indians also thought of the world's size in larger terms than the Mediterraneans, who did not imagine the existence of anything beyond the lands they knew, the oceans that surrounded them, and the heavenly crystal spheres that arched over their small pancake-flat world.

With Christianity's victories, this Judeo-Christian picture of a miniature world became a holy dogma, and the center of medieval thought. But although Christianity stifled free thought in western Europe during the Dark Ages, Islam proved to be more tolerant, and the philosophers of the Near East administered the Egyptian-Greek tradition well, supplementing it with Eastern sciences, especially those from India. Occidental astronomy is largely inherited from the Alexandrian scientists, and as a result most star names—for example, Aldebaran, Altair, Vega—are Arabic, as is the word "almanac."

But the great expansion in the Western man's perspective of his world that has taken place since the Renaissance was due not only to Arabic learning, but also to a group of artisans in Holland who taught themselves to grind lenses that could be assembled into a telescope. It was because of the cooperation between men who worked with their hands and thinkers —between the glassgrinders and the speculating philosopher-astronomers—that this great intellectual revolution took place. As the quality of telescopes improved, increasingly remote stars were discovered, and observations were made with increasing accuracy.

And so during the sixteenth and seventeenth centuries the narrow medieval world burst, and the earth, mankind's home and the center of creation, was dethroned to the status of one small planet, which, with eight other planets, revolved around the sun, the ruler of the planetary system. But during the eighteenth and nineteenth centuries a new revolution occurred in which even the sun lost its central position within the world order to become one star among hundreds of billions of others in our gigantic star system, our galaxy, which is 100,000 light-years in diameter. The Mecca of the universe, the center of the galaxy, became lost amidst the brilliant regions of the Milky Way in the constellation of the Archer (Sagittarius).

But the heavens were found to include a number of small, dimly shining points which disturbed the astronomers, and when the skillful glassgrinders of the twentieth century had made lenses and mirrors for today's giant telescopes, such dim points were discovered by the millions. Each feeble light source proved to be a galaxy, as enormous as our own Milky Way system. Thus occurred the third revolution in our thinking: even our own gigantic galaxy was too restricted for the searching human spirit, and our perspective of the world was shattered again. Today we know of billions of galaxies whose distances from ours are measured in billions of light-years, and we are discovering a yet larger system. Some believe that these galaxies constitute the entire universe, but others, who are more cautious, call it the "metagalactic system" and thus allow for the possibility that there may be other metagalaxies beyond our own.

With the shattering of the biblical conception of the creation, then, it was left to science to provide a new account. But since none of us was present

> *When the Eternal One sat*
> *and in the dark blue night*
> *sowed flaming seed*

we must proceed by guesswork. Our advantage today, however, is that our current guesses are founded upon a detailed modern scientific knowledge of the world's actual structure and upon a greater ability to reconstruct a picture of how the world looked a very long time ago.

To begin with, the biblical version has been replaced by *two* scientific accounts: that of the "little creation," concerning

The Cosmic Perspective

the earth's origin in one minuscule part of the universe; and that of the "great creation," or the creation of this enormous universe—that is, if indeed it *was* created.

Let us first discuss the little creation, or what scientists commonly call the *cosmogonic problem:* how the earth and the other eight planets of this solar system came into existence. This was, if we consider the enormity of the universe, a fairly minor occurrence, but for us who make our home on this mere kernel of sand, it is of the greatest importance. The first man to propose a scientific account of the earth's origin was the French philosopher and mathematician Laplace who, at the end of the eighteenth century, expressed his belief that the sun had once been condensed from a nebula, a very thin gas cloud. He suggested that not all of the matter in the cloud had been concentrated into the sun and that parts of the mass had halted at varying distances from the sun and were later condensed into planets. This theory has been doubted by many, and other possibilities have been discussed: one such alternative is that the planets were ejected from the sun by means of different catastrophes. However, the greatly expanded knowledge we have today of the various physical processes that can take place in a gas cloud has lent strong support to Laplace's general train of thought, although his ideas have been developed and modified in many respects throughout the years. It is likely, then, that electromagnetic phenomena were of decisive importance during the formation of the solar system. (In a subsequent section we shall discuss these phenomena more thoroughly.) By using different methods we have come to the conclusion that the event took place about four or five billion years ago—almost exactly one Kalpa ago, to speak the language of the Indian mythologers.

After the earth's formation, life eventually appeared on its surface, and in the course of millions of years higher forms evolved. "The earth brought forth grass and herbs bearing seed, each after his own kind, and the fruit tree yielding fruit after his kind, whose seed is in itself"—and then there were brought forth "great whales and all kinds of living and creeping things." Finally there appeared a very complicated being who called himself "man" and who regarded himself as the "crown of creation." One of his most remarkable accomplishments was his understanding of nature and his mastery of the natural forces. In this way he acquired for himself enormous powers. At the beginning of his existence he had been compelled by the force of gravity to crawl on the ground; but one day he conquered gravity and thrust himself out into space to conquer the cosmos. This noteworthy day occurred during the year he called 1957.

The "great creation" is called the *cosmological problem*. The first question we must ask is if the universe actually was created. Some think that there is an eternal god, like those of the ancient mythologies, who, at a certain point in time, created the universe. But the god who was equal to such a gigantic task can scarcely bear much resemblance to the minor thunder god who at one time was worshiped by a small tribe inhabiting one small area of one planet orbiting around a little sun, an average star among hundreds of billions of stars in a little galaxy, at least ten billions of which were mass-produced in the great galaxy factory of the creation. Others believe that it is just as simple, or perhaps simpler, to assume that the universe itself is eternal. Many good arguments—and many bad ones as well—have been brought forth in support of both ideas.

With the passage of each year, mankind's ability to explore the universe increases as does his brain's ability to interpret the instruments' results. The more knowledge we acquire about the structure of the universe, the more profound are the conclusions we can formulate about its early history. In this way we draw closer and closer to the question of its origin, and the problem of creation. But will we ever understand completely the mystery of creation—and was there a creation?

GALAXIES AND STARS

One of the most complex problems in physics is that on the age of the universe: is it infinitely old or did it come into being at a distinct point in time? Since this cosmological topic has been treated in one of my previous books (*Worlds-Antiworlds*), I shall not discuss it here. Let us examine, instead, a complex of problems of an entirely different character—those concerned with how our part of the universe got its present structure. If we assume that at one time our part of space contained a certain amount of matter, we then ask ourselves how this matter collected into galaxies. Why does a part of the matter within our galaxy exist in the form of extremely thin gases that fill a large part of the galaxy, while other matter is condensed into stars? Finally, why has at least one of these stars, which we call the sun, been surrounded with planets, among them Earth?

This complex of problems is of a type that is prevalent in natural science. We know the present state of things pretty well; we assume that the usual natural laws apply; and this assumption seems reasonable. We attempt to reconstruct the point of departure, about which we can make somewhat reasonable assumptions, and we try as much as possible to understand in detail how the process took place. The general character of the problem is similar to that of the topic on the origin and development of life, which was treated in the second chapter, "The Long Chain of Complications": we know about the present biological state, the organic life on earth, and the laws of biological development; and we are in a position to make certain general assumptions about the original conditions, that is, the possible chemical processes that took place on the earth before life existed.

Of the natural forces that have contributed to the present shape of our world, the force of gravity, gravitation, is the most important. All matter attracts all other matter, but the force of attraction decreases rapidly as the distance between the bodies of matter is increased. The force of gravity tends to make matter cohere. If, for one reason or another, there is a large cloud of thin gas in space, gravitation draws all the matter in the cloud toward its center, where it aggregates. All the matter in the cloud might quickly be concentrated into a giant mass if this process were not counteracted by the heat that the concentration generates. The process depends upon whether antimatter plays a decisive role. (The possibility that antimatter has decisive significance in this connection has been mentioned in Chapter II; see the footnote on page 27.)

If the concentration that has been caused by gravitation takes place within a relatively small volume of space, a star can be formed. If this star is large enough, the radiation emanating from it can produce a repelling force that prevents more gas from approaching the star.

If, in several different parts of the original cloud, there are areas in which the density becomes somewhat higher than that of the immediate environment of the cloud, a number of local condensations can form instead, so that the matter coheres into several different bodies. These attract each other but do not necessarily unite to form one enormous body. Instead, as a rule, they come to revolve around each other, and the centrifugal force of this motion counteracts the gravitational tendency toward further concentration. It can be assumed that a star system, such as a galaxy, is formed in this way.

As we mentioned in the first chapter, classical mechanics is the study of the motion of bodies, and general gravitation is one of the most influential forces in all such motions. With a knowledge of classical mechanics, then, we can understand at least in general how stars and star systems formed.

We know that they are essentially products of gravitation that have been equalized by the generation of heat and by centrifugal force. If we assume that in an original state the matter in the universe, or in a large part of it, consisted of a very thin gas, it is reasonable to think that this gas cohered so that certain condensations took place, causing galaxies to form. Within the galaxies a further condensation into stars occurred.

But for a more precise description and a more detailed understanding of the development of the stars and the systems of stars, we must look beyond classical mechanics and into other branches of physics, including nuclear physics. The enormous energy emitted by the stars is produced by nuclear processes taking place within the stars. The same forces that are liberated by an exploding hydrogen bomb produce, within the star, the energy that enables it to emit light and heat for millions or billions of years. Therefore nuclear physics is most vital to our understanding of the development of the universe.

Nuclear physics may also provide us with answers to the fascinating problem of the origin of the different elements. It is possible that the elements synthesized at one time from protons and neutrons, and there are many conjectures on how, when, and where the synthesis took place. Perhaps it was a result of the gigantic star explosions—the supernovae—which we observe occasionally, or perhaps it took place when the universe was in an early state of development.

PLANETS AND SATELLITES

We can assume that the sun was formed by the process discussed in the previous section. A body of gas, having dimensions hundreds or thousands of times greater than those of our present planetary system, began to condense, and gravitation drew the greatest part of the mass together to form the sun. The heat that was generated as a consequence of this process made the temperature in the interior of the sun rise to tens of millions of degrees. When the heat reached a certain intensity, the sun's immense furnace was ignited and nuclear energy ("thermonuclear energy") began to be liberated. The sun contains enough hydrogen to keep the furnace burning for billions of years, so that it can emit lavish amounts of heat and light into space.

If an astronomer on a planet revolving around a distant star were to observe the sun and its environment, the preceding information is approximately what he would consider worth noting. Presumably, he would not easily discover a detail such as that the sun is orbited by planets—as well as several thousand asteroids. It is even less likely that he would notice that many of the planets are, in turn, orbited by satellites. But for those inhabiting one of these planets, these details are not without interest.

How were the planets and their satellites formed? This question on the origin of the solar system is one of the most fascinating of the astrophysical classical problems. It was opened for discussion by the Frenchman Laplace several hundred years ago. Laplace proposed that when the sun was being condensed, its rotation caused a number of rings to detach themselves from the solar equator, and these later formed planets. Many weighty objections were put forth in opposition to this theory (commonly known as the Kant-Laplace theory). After subsequent calculations had shown that a detachment of the type outlined by Laplace could not occur, a number of completely different explanations of the origin of the solar system were attempted. One was the collision theory, put forth several decades ago by the English astronomers Jeans, Jeffreys, and others. They proposed that at one time the sun had collided with another star, causing the emission of a stream of gas, which later condensed into planets. However, because the distances between the stars are so vast, it is highly improbable that such a collision could occur. A more detailed analysis showed other flaws in the theory. For one thing, because the planetary system is as regular as it is, it is impossible that the planets are merely the splinters remaining after a cosmic smash-up. In addition, the largest planets, Jupiter, Saturn, and Uranus, are surrounded by satellites that form very regular systems of the same general type as that of the planets. This finding would force upon us the even more unreasonable assumption that these bodies, too, were produced by collisions.

It is probable that the same kind of process that produced the planets after the sun had been formed then worked "in minature" to produce the satellites of the largest planets. In other words, every large astronomical body tends to surround itself with smaller bodies: the sun has surrounded itself with a system of planets; and the largest planets, with a system of satellites. This realization returns us to a line of thought similar to that of Laplace, although once again we are confronted by the same weaknesses that were attributed to his theory.

The motions of the planets and the satellites within the present solar system are determined almost exclusively by the laws of classical mechanics. Both the Laplace theory and its derivatives have been based on the concept that during the formation, too, of the planetary system only mechanical forces were of significance. It is likely that this assumption is the root of the weaknesses in these theories. We have learned that when the planetary system was formed, the matter around the sun was in a gaseous form and the gas was ionized, that is, it was capable of conducting electricity. According to what we know now about the properties of such an ionized gas (a plasma), it is strongly affected by electromagnetic forces. It is therefore probable that electromagnetic forces were of decisive influence in the birth of the solar system.

If we use this concept as our point of departure, we see Laplace's theory from a different perspective, and as a result we can envision the origin of the solar system in somewhat the following way. After the sun was formed by the concentration of an enormous gas cloud, small parts of the cloud remained at a very great distance from the sun. Gravitation began to draw this remaining gas toward the newly formed star; but the sun's magnetic field halted the falling gas at varying distances from the sun—those at which the planets are now located. The forces of gravity and electromagnetism, then, effected a concentration and a condensation of the falling gas, and the result was the formation of the planets. When the largest planets had been formed, the same process was repeated on a smaller scale, giving rise to the systems of satellites. Thus the planet we inhabit was created by two great demiurges: gravitation, which forced the material in toward the glowing sun, and electromagnetism, which stopped it in space at the right distance. These same primeval forces also formed the features that enable us to inhabit this one planet—the mountains, water, and air.

THE ORIGIN OF THE MOON

The earth's closest neighbor in space is the moon. In attempting to determine how it originated, we must first ask if it could have been created in the same way as the moons of the three largest planets, Jupiter, Saturn, and Uranus, were: by the astrophysical process described in the preceding section. The answer is a definitive *no,* the reason being that the earth is too small and the moon too large. The largest of Jupiter's moons, for example, has only one ten-thousandth of Jupiter's mass, whereas the moon's mass is more than one one-hundredth the mass of the earth. The process by which the giant planets acquired moons probably cannot even take place around bodies as small as the earth. The earth and moon, then, are not to be likened to a planet and its satellite, but rather they can be thought of as a double planet, even if the two are not equal in mass.

Our knowledge of the tides has been most important to our understanding of the moon's history. Twice each day the moon makes the ocean waters rise and then recede again. But according to the laws governing action and reaction, the flood and ebb tides also affect the moon and change its orbit. The effect is very small, but in the course of millions of years the tides cause the moon's distance from the earth to increase, and simultaneously they slow down the earth's rotation. The day is thus lengthened—but only by a few thousandths of a second each century.

If the moon is now receding from the earth, clearly it must have been closer in the past. It is therefore probable that a long time ago—perhaps a billion years—the moon was very close to the earth. Did the earth at one time eject our steadfast old moon, so that the Russians and Americans were not the first, after all, to launch their miniature moons out into space?

About one hundred years ago the English astronomer Darwin put forth the hypothesis that before the moon existed the sun caused enormous tidal waves on the earth. These waves became so immense that a large part of the earth's mass was expelled and later formed the moon. Others later elaborated upon this hypothesis by proposing that the moon left behind a cavity, which we call the Pacific Ocean.

Although this hypothesis has been widely disseminated

through irresponsible popular works, it was refuted a long time ago and today is not taken seriously by professional astronomers.

We have been sure for a long time that the moon could not have been ejected from the earth in any way, because of the unusually high amount of energy that would have been required to expel a mass that large. But if we calculate changes that have taken place in the moon's orbit we find that once the moon must have been very near the earth. Where, then, did it come from?

To answer this question, let us consider the four innermost planets, Mercury, Venus, Earth, and Mars. At first glance, they appear to form a homogeneous family of small planets separate from the large planets, Jupiter, Saturn, Uranus, and Neptune. But if we study the small planets more carefully, we find that Mars does not conform to the system that exists among the other three. First of all, we would expect the fourth planet of this family to be much larger than the earth; but the mass of Mars is only one-tenth of the earth's. Furthermore, Mars has a density as low as 4.1, whereas the other three have densities between 5 and 6. If we look around for the close relatives of Mars, we find that the moon is a likely possibility. Like Mars, the moon has a low density (3.3), and its mass, which is about one-tenth that of Mars, is what would be expected if the moon were a planet orbiting between Mars and the earth. As a result, it is believed that the moon may have originated as a planet at the same time as Mars, and that it moved in an orbit that occasionally brought it very close to the earth. The earth, then, would have been formed independently of both Mars and the moon, and is actually a member of a *three*-planet family—Mercury, Venus, and Earth.

On this basis, it is reasonable to conclude that the earth by some accident captured the moon, demoting it from an independent planet to a satellite. However, although this theory is an attractive one, there is one substantial difficulty: because a captured body would necessarily orbit at a great distance from the earth, the moon, if it is indeed a captured planet, must have moved in a wide circle around the earth; yet the tidal calculations indicate that the moon orbited very close to the earth during the earth's very ancient history.

We may resolve this difficulty by studying the calculations made by a German astronomer, Gerstenkorn, on the tidal effect. With the highest precision, he has calculated the changes that have taken place in the lunar orbit. The moon now revolves around the earth at a distance of 60 earth radii, and its direction of motion is the same as that of earth's rotation (direct motion). Its orbit is inclined on the average of 23 degrees with respect to the earth's equatorial plane. Gerstenkorn substantiated the finding that about one billion years ago the moon had orbited at a much shorter distance from the earth, but he also learned that at that time the orbit was inclined at an even greater angle to the equatorial plane. At an even earlier time, the lunar orbit's angle of inclination was so great that the moon revolved directly over the earth's poles; and still further back in time the moon moved in a direction opposite to that of the earth's rotation (retrograde motion), and the tidal effect worked in a direction opposite to that of the present age. When, therefore, Gerstenkorn continued his calculations back into time, he found that at one point the moon must have moved in an orbit at such a large distance from the earth that it was an independent planet.

It is possible that this theory has provided us with the solution to the moon's ancient history; it has also supplied clues to a number of other problems. The thicket of mathematical formulas has yielded accounts of a series of catastrophic and dramatic events in the ancient histories of the earth and the moon.

The moon, originally an independent planet moving close

to the earth's orbit, accidentally came so close to the earth that it was captured, so that it began to circle the earth instead of the sun. But because its orbit was retrograde—that is, its motion was in a direction opposite to that of the earth's rotation—the tidal effect worked in a direction opposite to that of today. The lunar orbit continually diminished but at the same time its angle of inclination to the equatorial plane continued to increase. Finally the moon revolved directly above the earth's poles. At that time it was at a distance of only about 3 earth radii from the earth, and its orbit continued to diminish.

Had there been people on earth to observe the moon and its changes, they would first have seen the moon shining as a planet, looking much as Venus or Jupiter do today. After its capture the moon would have appeared somewhat as it does now. But to the observer, it would have appeared, in the course of millions of years, to be growing larger as it gradually approached the earth until finally it looked like a disk with a diameter twenty times as great as that of the full moon visible now. As the moon drew closer to the earth, the tidal force grew, and when the moon was at the point nearest the earth, the flood wave had a height of more than a mile. But if the moon produced enormously large tides on the earth, the tidal effect of the massive earth upon the small moon was much greater. The effect of earth's gravitation on the moon's surface became so strong that it actually exceeded the moon's own gravitation. It has been known for a long time that this is exactly what occurs as soon as a satellite begins to orbit inside the so-called *Roche limit,* which is a distance of about 2.9 earth radii.

According to Gerstenkorn's calculations, the moon did come within the Roche limit, at which time it began to break apart. Fragments of varying size, such as stones and gravel, were drawn out from the moon by the earth's force of attraction and filled the entire volume of space around the earth and the moon. But once the moon had penetrated the Roche limit, it had reached its minimal distance from the earth, and it slowly began to recede. Some of the moon fragments may have fallen onto the earth. Some fragments fell gradually back onto the moon, and it is possible that the lunar craters were produced in this way. We do know that the moon's present scarred appearance is the result of its brutal encounter with earth's superior force of gravity. But once it had receded from the Roche limit, its only other significant change was the increase in its distance from the earth as a consequence of the tidal effect.

And how did this catastrophic encounter change the earth? The enormous tidal wave that swept the earth for a long period of time must have "polished" it very thoroughly. Afterward, some of the moon fragments fell to the earth, but we cannot determine how large a quantity that was: it is possible that the amount is insignificant, and that it would be useless to look for such remnants. Gerstenkorn's theory has provoked much discussion, and geologists are trying to find traces of the events in the geological history of the earth. Evidence of a geological catastrophe that took place 700 million years ago suggests that the moon approached nearest to the earth at that time. However, much more research is necessary before we can confirm the Gerstenkorn theory and reconstruct the event with certainty.

ARE WE ALONE IN THE UNIVERSE?

Artificial satellites and deep space probes have stimulated man's interest in the inveterate question on the possible existence of life on other worlds in the universe. Within our own solar system, Mars is the only other body that may possibly have the conditions for organic life. But since the tundras of northern

Siberia and the peaks of the Himalayas have climates that are pleasant in comparison to the Martian climate, it is hardly believable that the organic life—if any—can have attained higher development. "Men from Mars" appear in science fiction, but, without a doubt, nowhere else.

Although it is unlikely that there are, within our solar system, beings who have established a civilization in some way comparable to our own, we cannot eliminate the possibility that such a civilization may exist on a planet orbiting another star. This can, of course, be nothing more than loose speculation, but speculations are far from uninteresting.

Underlying this topic is the question whether stars other than the sun are surrounded by planets. We cannot answer this with any certainty. To demonstrate the difficulty in acquiring accurate knowledge of this, let us imagine a model of our solar system in which the sun is represented by an orange and the earth by a grain of sand ten yards from the orange; the distance between this orange and another one, representing the nearest star, is then equivalent to the distance between France and New York. To determine if a star is surrounded by planets is as difficult as discerning from an observation point in New York a dark grain of sand placed ten yards from an orange in France—a task that could not be accomplished with the largest astronomical instruments available.

Opinions about the existence of other planetary systems have changed concurrently with our conception of the origin of our own planetary system. Recall that the proponents of the collision theory believed that our system of planets came into existence when the sun collided with another star several billion years ago. Since the probability that one star would collide with another is as small as the probability that an orange ejected in an arbitrary direction from New York would collide with an orange in France, it could be concluded that in all likelihood only a few of the hundreds of billions of stars in our galaxy could have managed to collide in this way. Thus, at the time the collision theory was prevalent, it was thought that the number of planetary systems like ours must be very few indeed, and that those that did exist must be in completely different sectors of the galaxy.

But the collision theory has since been replaced by the more probable supposition that our planetary system resulted from a process that is directly connected with the origin of the sun. Although we are still a long way from agreement on the details of the origin of the planetary system, the predominant assumption is that it was the result not of an exceptional celestial occurrence, but of a perfectly normal process connected with the formation of a star. It therefore follows that many stars—perhaps a majority of them—may be surrounded by planets having approximately the same characteristics as the planets revolving around our sun. This means that there is a high probability that somewhere there is a planet, having about the same physical and chemical characteristics as the earth, that revolves around a star the size of the sun.

The next question is if it is probable that a form of organic life appears on such a planet. Arrhenius' opinion that life can spread throughout space by means of "spores" does not have many disciples today. Rather, as we discussed in Chapter II, it is generally believed—as the Russian scientist Oparin and others have proposed—that the simplest living systems on the earth were a result of the formation of very complicated carbon compounds from inorganic matter under the influence of solar radiation and other phenomena. This view maintains, then, that the origin of life was quite a normal occurrence under those conditions which prevailed on the earth when, several billion years ago, the simplest living beings appeared. Furthermore it is not impossible that life might have originated under conditions such as those on Mars, and it is quite probable that it exists on a planet orbiting another star if that planet has somewhat the same physical and chemical structure as the earth.

If these theories are correct, we should assume that of the hundreds of billions of stars in our galaxy, several billion—or at least several tens or hundreds of millions—are circled by planets bearing living organisms on their surfaces. Has life on any of these planets developed in a way similar to the way it did on earth? Has the long chain of complications functioned in the same way? Has the result been a form that resembles man? And, if so, will we be able to make any kind of contact with these beings?

Because the development of rocket technology has enabled us to build spaceships capable of reaching Mars, in the foreseeable future we can expect to ascertain whether life exists there. At present it does not seem likely. Certainly Venus is not inhabited. Trips to other planets within the solar system will also become possible within the not too distant future; but the prospect of finding higher forms of life within our planetary system is very small. A journey to an earth-like planet of another solar system is, because of the enormous distance that would have to be traversed, so difficult that at present we do not see how it can be done. A space ship sent out of our solar system with the speed of the present satellites or deep space probes would require close to 100,000 years to reach the nearest star. Even if the speed could be multiplied by 100—which would be an enormous technical achievement—the journey would take 1,000 years. There is therefore no imminent possibility that, by the application of any presently known phenomena, we might send an unmanned spaceship to investigate the conditions around the nearest stars.

There is a possibility of ascertaining the existence of life on planets in distant solar systems only if such life has evolved into intelligent beings who have developed highly advanced radio or rocket technologies. It is theoretically possible that a radio transmitter on a planet of another solar system in our region of space might reach us with audible signals. The necessary transmitter would have to be powerful, but not unreasonably so, provided that the aim of the transmitter's radio beam were directly at our solar system. Speculations have been made on what wavelength these hypothetical beings would choose, and what kinds of signals they would send if they believed that we existed and had a sufficiently developed radio technology to receive the signals. If we did receive such signals and built a giant radio transmitter with which we could answer, a two-way correspondence could be established, although this would be difficult. Since the radio waves move at the velocity of light, a message would require four years to reach the nearest star. Thus it would be at least eight years before we received an answer to a telegram we had sent to a planet orbiting this star. It is, however, most improbable that one of our closest neighbors would be ready to make contact with us. A more likely correspondence would be one across a distance of, say, 100 light-years; for an answer from that planet we would have to wait 200 years. If, as a result of the development of a highly acute intelligence, we could then understand what the signals meant, a most interesting exchange of telegrams could take place.

Another possibility for contact would arise if the inhabitants of the distant planet had already advanced so far in rocket technology that they could send a space vehicle with which we could communicate at close range. This kind of speculation underlies the stories about "flying saucers."

Since we have yet to hear any radio signals from distant planets or to observe any spaceships from them, we have absolutely no indications that distant civilizations exist. Whether we should believe in their existence is a question for the biologists, sociologists, and historians to answer. If we presuppose that there are many inhabitable planets in our galaxy, what is the probability (1) of life appearing on them and (2) of it then developing in such a way that an organism as complicated as man could arise? That is a question no biologist can answer, but it might perhaps be elucidated by a detailed study of biological development. Did pure chance determine each of the many stages in this development: the appearance of the simplest living aggregates, of the cells, of the multicellular beings, of their increasingly complicated descendants, and, finally, of man? Or did biological development move toward higher complication in an inevitable and regular way? And once man had evolved and his societies had formed, when did a scientific and technical culture—one that could measure the distances to the stars and begin to speculate on their conquest—become necessary? Was it all a chain of accidents so unique that in all probability it could have happened only at a single location, namely here on earth, and nowhere else in the galaxy? We do not know enough about the ways in which society functions and develops to determine if we are unique in the universe.

The conjecture that we may at one time be able to communicate with sophisticated beings on another astronomical body is so enticing that it is difficult to divert our attention from it. Certainly we shall continue to hear frequent press reports about radio signals and saucers from afar, but the probability that they are true is very small. If remote cultures do indeed exist, the obstacles to establishing communication with them are so great that in the foreseeable future the only descriptions of life on distant worlds we can expect are those supplied by fanciful authors.

V

Natural Science and History

In recent decades an extraordinarily dynamic development has taken place in natural science and technology. Within less than a quarter of a century we have entered a number of new "ages": the atomic age, which revolutionized warfare and world politics; the computer age, which is in the process of effecting an organized revolution within society; and—perhaps most significant of all—the space age, which has enabled man to leave the earth for the first time. In comparison with the slow development of the preceding fifty or more centuries that we have some historical knowledge of, that of the twentieth century, and especially the past twenty-five years, seems rapid indeed. Many humanistically inclined people, who believe that the former rate of change is the normal one for human society, are disturbed and perhaps enraged at the extremely rapid pace that natural science and technology have introduced.

But natural science also provides a totally different outlook on this rate of development, namely, the long-range view on man and his culture that is obtained from a geological or cosmological perspective. In these scientific fields the duration of a phenomenon is not measured in years or decades or centuries; the time scale is rather millions or billions of years. It is interesting to try to place human development, both the gradual historical evolution and the rapid progress we are experiencing now, in the geological-cosmological context. But it is at first difficult to comprehend periods of time as long as those involved in these fields. Although we can talk about millions or billions of years, few of us have any real understanding of what this means. A thousand years is a long time, a million years is certainly longer, and a billion years is longer still, but the proportions of these time periods to one another lie beyond man's everyday experience.

Let us therefore make use of a reduced time scale so that we may relate man's development to geological development: assume that a second represents 100 years. The earth, then, originated as a result of the cosmogonic processes somewhat more than one year ago. Life appeared on earth several months ago. The transition from ape to human being took place one or two hours ago. Since history tells us that human cultural development has taken place in the course of the past 6,000 years, on our reduced time scale human history began one minute ago. The industrial revolution occurred during the last second, and the atomic age, the computer age, and the space age all began during the last few tenths of the past second.

In the earth's ages-long historical development, the "minute" we have just experienced is therefore something exceedingly dynamic and remarkable. From the cosmological-geological perspective, the entire course of human history has thus proceeded at a furious rate. The transition from primitive agriculture to our own time can be likened to a single explosion. Everything described by history has taken place during the last minute of the earth's year-long existence. We might ask if at any time during the earth's previous history there has taken place an event as dramatic as human history.

It is conceivable that one or two times before, similarly rapid changes occurred. For example, rapid geological changes might have resulted from the enormous tidal waves that, according to certain calculations, were caused by the moon's close proximity to the earth. But this topic is beyond the scope of our present discussion. The other, more important event that may have been as explosive as human history is the appearance of life. When the earth originated, it was sterile. In the course of many hundreds of millions of years, increasingly complicated chemical compounds were produced. Finally the molecular aggregates attained a degree of complication that enabled them to reproduce themselves and to grow by taking nutriment from their environment. It is possible that this produced a very sudden development. Had the first organisms possessed reproductive powers as enormous as those of microorganisms existing today, they would have been able to cover the earth within a very short time. It is probable that their reproductive capacity was less; yet within a short time, perhaps within "minutes" on our reduced scale, primitive life dominated a part of the earth—probably the seas in particular—that could harbor life forms. Once this explosion of life had occurred, the development was more gradual. The processes of differentiation and selection that took place in the subsequent geological ages produced an increasing number of species until man appeared, initiating on the earth a new and dramatic era.

A question about the next stage of our development now arises. What will happen during the next "second," the next "minute," and the next "hour"?

The dominant factor at present is the development of science and technology, which have produced entirely new conditions for mankind. Thanks to modern technology, the earth can feed a much larger population at a much higher standard of living than ever before. Medical science has advanced so far that it is possible to stop, without serious difficulties, the population explosion which necessarily reduces the standard of living, no matter how rapidly modern technology develops means to raise it. We have the scientific and technological ability to regulate the world's population to an optimal one, and to allow this population to enjoy very high living standards. The only obstacle to the realization of this happy vision is the chaotic situation prevailing in world politics. Many people suffer serious want, not because we can*not* feed them, but because too many of the world's politicians are incompetent. We shall not discuss here the possible political solutions to these problems, but let us consider some of the effects of natural science and technology on the world today and some of the alternatives that these two fields might offer to the current political impasse.

The dominant feature of the present age is the exponential growth of human knowledge and skill. Within a short period of time—let us say a few seconds on our reduced time scale— our rapidly expanding science and technology will allow us to fulfill many wishes. The development of the communication media has caused the world to contract; in fact, some maintain that its size is diminished so much that it is perhaps in the process of becoming too small for human technology; certainly it is true that science and technology are transforming the earth at a constantly increasing rate. Its natural resources are being consumed, and some of them are being exhausted. Air and water are being polluted; in the arsenals of the superpowers there are enough atomic bombs to pollute the entire planet. Soon it will be possible to change climates. The composition of the air has changed noticeably since the beginning of the industrial revolution, and it can be changed even further. Thus it is easy to believe the earth is indeed becoming too small for the science and technology of the future.

As a result, the beginning of the space age may be the most important event to have occurred in human history (with the possible exception of the beginning of the computer age). Man can now free himself from the earth that is too small to accommodate his powers of creation, and he can take himself into the space that surrounds him. The question is, what will make him do it?

Since the launching of the first space vehicles opened up new horizons to us, research has been devoted to the proper conditions for space flight and the electromagnetic conditions in the region of space immediately surrounding us. Lunar landings are being prepared, and astronomers are transferring observatories from the earth to space ships and the moon, where they can be operated without atmospheric interference. Occasional space vehicles have reached Venus and Mars to make a few observations of these bodies at close range. But what are the possibilities for continued space activity? What will happen in the next few "seconds" and "minutes"?

The answer to this depends largely upon how much fantasy and power of action man is capable of developing. If these should be restricted by political conditions, little advancement will be made in the field of space. That a constantly expanding technology is being confined to a planet that is already too small for it manifests, perhaps more than anything else, the destructive aspect that inevitably shows itself in a power of this magnitude if it is constrained within an area that has become too small for it. Perhaps a catastrophe can be avoided only if man has the foresight and imagination to let technology expand, and himself with it, out into space.

What are the possibilities for the exploitation and colonization of space? Among the astronomical bodies that are closest to us, the moon is not a primary possibility for habitation, because it lacks an atmosphere and has a very severe climate. The atmosphere of Mars is too thin to sustain human life. That of Venus is too dense, and its chemical composition is such that, in its present state, it is no better suited than the atmosphere of Mars for human life.

But note the phrase *in its present state*. At the time that life appeared on the earth, this planet too was "uninhabitable." Its atmosphere was perhaps similar to that surrounding Venus today. It was probably composed chiefly of carbon dioxide, and because it contained very little oxygen, the "higher" forms of life could not exist there. But when life began, it changed the earth's living conditions itself. The earth has become inhabitable for higher life only because of the ability of primitive life to transform earthly conditions. This transformation was one of the most important results of the first explosion in the history of life—the explosion of life itself. What will be the result of the second explosion—that of human technology?

We know that life multiplied expansively, once the molecular aggregates reached a sufficient degree of complication. Now men have learned to cooperate with each other and with the machines that they themselves have created. The integration of computers into society has been of decisive significance. The technological explosion is in the process of changing the entire earth, making it more habitable in some respects and less so in others. If technology can thus revolutionize one entire planet, the earth, it will soon be able to transform others. Although Mars and Venus are not inhabitable, one worthwhile task for an expanding technology would be to make them so. Our microorganism ancestors transformed the earth. Why should not we be able to transform—perhaps with the assistance of microorganisms—the neighboring planets into pleasant locations to accommodate our growing race?

This is perhaps one of the most important events that will take place in a not too distant future. The "seconds" we are now experiencing are thus a preparation for the "minutes," or perhaps "hours," during which the life that appeared on the earth will begin its cosmic expansion.

THE PHILOSOPHY OF NATURAL SCIENCE

As a conclusion to this survey, we shall consider the world-view of natural science and the relationship of that view to religion, knowing perfectly well that this topic is much too comprehensive for this short essay (although a similar remark can be made about every section in this little book!).

Undoubtedly, some controversy has existed between religion and science for a long period of time. When the progressive nature of science became too prominent, the protectors of religion would try to retard its growth by censorship and persecution. The Inquisition realized quite correctly that many of Christianity's most cherished dogmas would be seriously threatened if the critical spirit, which is the very life of science, were not extinguished. Similarly, many of the defenders of natural science have been sharply critical of religion. It is nevertheless indisputable that many outstanding natural scientists have been deeply religious. Certainly some religious statements by scientists—especially in earlier times—can be explained as the result of a desire to avoid what could be a truly perilous conflict with the ecclesiastical authorities; other such statements have been made because many—perhaps the majority—of those in science are so engrossed in their special area of research that they have never given serious thought to these profound issues and therefore accept conventional religion uncritically. But, it is not impossible to accept a scientific account of life on earth and at the same time to be religious, as long as the word "religion" is not given in its narrow and conventional meaning. In fact, it is scarcely possible to ponder the long chain of complications without feeling a certain religiously colored reverence for the miracle wrought by nature: a miracle that is all the more fascinating for not being a simple juggler's trick. Each link in the chain is small, simple, and self-evident—at least it is the aim of science to make it look that way—but it is the whole chain that constitutes the great miracle, from atom to man.

A God who personally and repeatedly intervenes in the course of events in order to favor his worshipers and punish his blasphemers is, of course, completely incompatible with scientific thought. Yet in those areas outside the limits of human observation, the possibility of a divine source is at least admissible. No matter how far back into time we manage to trace the "written" cosmological history, and no matter how deeply we grope toward nature's most fundamental laws, there is always something beyond. We are unable to give a scientific answer to the questions, how was the world created, and what was there before the most ancient events we know of? To reply, therefore, that the world was created by God, or by Brahma the Creator, is not to come into fatal conflict with science. Nor are we rejecting scientific thought if we say that the natural laws—or the most fundamental natural law— were established by God, and that the omnipotence and omnipresence that science ascribes to the natural laws can also be attributed to Vishnu the Preserver. But an atheistic outlook is at least as justified.

We have already briefly considered the question of the soul in Chapter III. This is one issue in which the dispute between religion and science becomes particularly acute. Is the soul, as idealists insist, of another nature than the body, descended from a higher spiritual world—from which it was banished after the Fall—to this one, where it is entangled in sin and a material existence for a while? Or is it something the brain secretes, much as the liver secretes bile, as the most extreme materialists have maintained?

The conflict between the two has been tinged by a considerable amount of emotion. The idealist is quick to label the materialist "crass," believing that his inability to "raise himself" to the idealistic concept is due to his incomprehension of all spiritual values. Life would lose its meaning for some idealists if they did not believe in a higher spiritual reality. The materialists have answered that religion is the opium of the people, consisting of myths invented and propagated by priests that in reality serve evil social forces.

Religion's spokesmen have often accused natural science of being materialistic, and this is largely correct. Particularly in its infancy, natural science worked almost exclusively with the common materials and mundane occurrences despised by idealists: the falling of stones, the flowing of liquids, chemical reactions. By the exercising of critical and constructive thought, and by the resultant improvement in their ability to observe and study the world methodically, scientists have become capable of examining phenomena of an increasingly intangible nature. The two significant breakthroughs were quantum mechanics and the theory of relativity. Quantum mechanics has shown us that matter consists of elementary particles, which from a certain standpoint must be considered wave motions. The theory of relativity has revealed that matter and energy are equivalent; matter, then, can be seen as a form of energy. These new observations should significantly alter the purely emotional reaction to materialism. It is no longer possible to call matter "common," or materialism "crass" because our concept of matter has now become more abstract—we might say more spiritual—than any religious concept of a God or other divine being. It is easy to understand why an idealist with an exalted conception of his own soul and its mysteries cannot accept the idea that these have a material basis, as long as he associates the word "matter" with something strictly tangible, a mechanical machine or an automaton. But his objection may subside if "matter" is defined as "a form of energy or a wave of motion." But why, after all, should we trouble to retain the word "materialism," marked as it is by so many conflicts? In truth, the basis for the old materialism was swept away when physics put forth a particularly "idealistic" conception of matter itself.

Whether the world-view of natural science is idealistic or materialistic is, consequently, an all but inessential question of terminology. What *is* of decisive importance is whether it is possible to achieve a unified world-view that is based on scientific fact. To this end, some resolution of the controversy on the nature of the soul is of fundamental importance.

Scientific analyses of mental phenomena are carried out within the field of nerve physiology, which is concerned with the functioning of the brain and the nervous system. In accordance with the scientific method, the physiologist's first step in such an analysis is to understand the very simplest phenomena, in order that he may acquire a foundation that he can use to analyze at least the general features of increasingly complicated phenomena. In an analysis of mental phenomena, then, we would begin with the reflexes of animal and human bodies. The research of the Russian biologist Pavlov on the reflexes of dogs has been especially revealing.

The most important reflexes are those necessary for the maintenance of life. If food is placed in front of a dog, secretion of its saliva and gastric juice begins immediately. The sight and smell of food produce a nerve stimulation that is conveyed to the salivary glands, where it causes increased secretion. The dog moves toward the food, and even this movement can be considered a complicated reflex. By having a bell ring every time food was displayed, Pavlov succeeded in producing a new type of reflex, the "conditioned reflex": if the bell rang, the dogs' secretion of saliva increased and they rushed toward the bowl of food even if it did not contain food. It is beyond the scope of our discussion to go into other, more complicated patterns of reflexes that can be elicited in this way, and into the strong similarity of these reactions to the simplest of the mental phenomena.

Pavlov's experiment has given us some understanding of at least the general physiology underlying a dog's mental processes. The mental processes of the human being are incomparably more complex than those of the dog; but is there really any reason to assume that something basically new has appeared? Is the difference between the reactions of a man and those of a dog greater than that between two links in the long chain of complications? Certainly there is an extremely remarkable and apparently unfathomable element in the human consciousness that governs man's attempts to explore and understand everything, including himself; but if we reflect on the past successes of science in finding unity in the most disparate phenomena, we cannot regard as unreasonable the possibility that someday even consciousness and the soul will be analyzed successfully and understood. Once this is accomplished, we will have the prerequisites for a unified world-view.

When the materialists have had occasion to liken the human mental processes to the functioning of a machine, the comparison has met with indignant protests from the idealists. Admittedly, if "machine" is defined merely as something that functions automatically according to a fixed schema, then to compare human mentality with a machine is indeed a misinterpretation. But the computer is an example of a machine that can function in an extraordinarily complicated way. Even though the machine may be constructed of simple elements, of which the function of each is known, the processes resulting from the coordination of all these elements exceed all human comprehension. When the computer is started, no one knows what it will produce as a result—it is used precisely for solving an unresolved problem. Thus, as we have already noted in Chapter III, it is possible to draw a striking analogy between the methods of a computer and those of the human brain in treating a problem. But of course significant differences do exist. While a mathematician is making a calculation, he may, in the middle of it, begin to hum a Beethoven sonata or to plan what he will have for dinner. Because a computer is never diverted in this way, it calculates much more reliably than the mathematician who allows himself to be distracted.

Perhaps the difference between a mathematical machine and a human brain is largely due to the greatly superior complexity of the brain. Its components, though simple, are extremely numerous, and they perform many different functions. For example, one of its many parts concentrates on the solution of mathematical problems (this part is missing in certain specimens!), another focuses on the mundane business of making a living, and another connects the first two parts so that they affect each other. Although the electrical impulses or waves that travel between the different parts are subject to the normal natural laws, the result is incomprehensibly complex. To select a new metaphor, we might compare the soul to an ocean in which wind and underwater currents produce waves. Even if the laws governing the wave motions and the wind's effect are known, and even if the contours of the beaches and the formations of the ocean floor have been thoroughly and precisely mapped, there is still no possibility of calculating in detail how and when a wave will break over a rock. We can state with certainty that a certain wind strength and a certain wind direction will produce waves having a certain average height, and that the breakers on a certain part of the beach will be especially dangerous. Similarly, we can generally predict the way in which a particular individual will react under certain external conditions. But no one can predict exactly how a wave will spray upward or how the sun will glitter on the sea at a given moment, not because we do not know the laws governing wave motion and the reflection of light, but simply because the problem is too complicated. And the human brain is at least as complicated as an ocean.

This analogy could be countered with the objection that atomic phenomena are much more important in the human body than in such a macroscopic occurrence as the movement of waves on the surface of an ocean. It is true that if the atomic phenomena are significant, there may be a basic limitation on the possibility of predicting what will happen. Heisenberg's widely discussed uncertainty principle maintains that it is not possible to determine exactly both the position and the velocity of an electron simultaneously, because each measurement that is made affects the phenomenon being studied; consequently, if we cannot know both the position and the velocity at a given moment, we cannot calculate in detail how the electron will move. The uncertainty principle has been used and misused in many problems, including the timeless question on freedom of the will. In physics, in which it was previously thought that the precision with which an experiment could be carried out was absolute, the uncertainty principle is now an important factor to be considered. But in biology, the conditions for observation have never been as precise as those of physics, because any phenomenon under study consists of so many variable components. Consequently, the objection is invalid: the uncertainty we might face as a consequence of the complicated nature of the problems does not, as a rule, increase to any noteworthy degree because of the extra uncertainty introduced by Heisenberg's principle. (The discussion on the eye's sensitivity gave an example of a phenomenon in which Heisenberg's principle is important.)

Before the advent of natural science, religion made a unified world-view possible. Now modern scientific thought has precluded this possibility. As long as it is believed that the soul belongs to God and that it obeys his laws, while the body and other parts of the natural world obey the natural laws, a dualism, a tragic conflict, is inescapable. Man comes into conflict with nature by asserting that he is a god, even though a fallen one. He does not want to admit that his ancestors were as humble as those of the amoeba, and he is especially unwilling to acknowledge the apes as cousins. Only if he abandons this arrogant attitude can man achieve the harmony, the unity with all else that exists, which is the goal of every world-view. He must acknowledge that he is part of nature; he must understand that the carbon atoms that constitute soot and diamonds can combine with others to form protein, urine, amoebae, lilies, or human beings: that we are all more or less accidental aggregates of atoms—that from dust we came and to dust we will return. Man, after all, constitutes a pattern of electron waves, as do all other objects. But this is not to say that he is only a conglomeration of atoms. We might just as well describe a given painting as consisting of 10 grams of yellow paint, 20 grams of red, and 30 grams of green. What makes the painting a painting, and perhaps a masterpiece, is the way in which the colors relate to each other. Art is not color but combination: and it is the combination of atoms and the ways in which they interact that constitute the living man.